QUANTUM PHYSICS

FOR BEGINNERS

The most compelling phenomena of quantum physics made easy:

the law attraction and the theory of relativity

[Brad Olsson]

Legal & Disclaimer

The information contained in this book and its contents is not designed to replace or take the place of any form of medical or professional advice; and is not meant to replace the need for independent medical, financial, legal or other professional advice or services, as may be required. The content and information in this book has been provided for educational and entertainment purposes only.

The content and information contained in this book has been compiled from sources deemed reliable, and it is accurate to the best of the Author's knowledge, information and belief. However, the Author cannot guarantee its accuracy and validity and cannot be held liable for any errors and/or omissions. Further, changes are periodically made to this book as and when needed. Where appropriate and/or necessary, you must consult a professional (including but not limited to your doctor, attorney, financial advisor or such other professional advisor) before using any of the suggested remedies, techniques, or information in this book.

Upon using the contents and information contained in this book, you agree to hold harmless the Author from and against any damages, costs, and expenses, including any legal fees potentially resulting from the application of any of the information provided by this book. This disclaimer applies to any loss, damages or injury caused by the use and application, whether directly or indirectly, of any advice or information presented, whether for breach of contract, tort, negligence, personal injury, criminal intent, or under any other cause of action.

You agree to accept all risks of using the information presented inside this book.

You agree that by continuing to read this book, where appropriate and/or necessary, you shall consult a professional (including but not limited to your doctor, attorney, or financial advisor or such other advisor as needed) before using any of the suggested remedies, techniques, or information in this book.

Table of Contents

Introduction

Everything we know consists of matter made up of particles, i.e. atoms, which consist of smaller particles: electrons, protons and neutrons (the last two are made up of quark particles), and invisible energy interactions or forces, among which are gravity, the attractive and repulsive forces of charged objects or magnets, X-rays, gamma rays and radio waves (mobile phones, Wi-Fi, radio and TV).

There are only four invisible forces, in fact. The first force, gravity, is vital for our environment as it ensures that all objects around us stay in place, including keeping each of us on the ground as well as keeping the Earth in consistent orbit around the Sun.

Yet, it is the least strong force of all. There are two much stronger forces inside atomic nuclei (called simply weak and strong interactions), which ensure the existence and stability of the nucleus of each atom in the known universe (including the atoms that make up each of us).

The rest of invisible reality (outside gravity and nuclear forces) is categorized into the subtypes of electromagnetic forces or electromagnetic radiation (each force always has a source, i.e. it is always radiated by something), including the above-mentioned radio waves, X-rays and gamma rays, as well as some others, as we will come to.

Now, to begin the study of quantum physics it is necessary to have a general understanding of electromagnetic radiation, as quantum physics was born out of attempts to gain an insight into it.

Chapter 1 : What is Quantum Physics

In addition, scientists use these mathematical equations to explain what they observe in the world around them and also what they observe through various experiments. As the tools of their trade have become more precise, scientists are able to gather better information to add to their understanding of the molecular world. Today, we benefit from the work of these scientists to better understand our world and the Universe on a molecular level. As we will see, Quantum Physics is mathematics at work explaining the world around us, down to the smallest detail.

These theories include wave particle duality and quantum tunnelling. Yet before the scientists could create these theories, there were plenty of experiments which assisted them in formulating these theories.

The world of mathematics

No doubt you learned physics at school. What do you remember? Probably that it's always associated with mathematics. Which doesn't make physics any more attractive. But unfortunately one cannot avoid mathematics, which is indispensable for physics. Galileo Galilei (1564 - 1642), the father of modern physics, put it this way: "The book of nature is written in the language of mathematics". Nevertheless, practically everything can be explained without it. Just like I do in this book. At least almost, because I will introduce a very simple formula.

The age-old question is whether mathematics is an invention of man or whether it has an independent existence. That's still controversial today. But most mathematicians and nearly all physicists tend to its independent existence. This idea goes back to Plato (428 BC - 348 BC). However, he did not speak of mathematics, but of the world of ideas and perfect forms. Which serve the universe as role models. Let us join the majority and assume the independent existence of the world of mathematics. So what is the universe? By definition, it represents the totality of what exists. Consequently, it encompasses two worlds. Once the material world, that is the world of matter and energy, embedded in space and time. The universe also includes the world of mathematics, to which we ascribe an independent existence, detached from us humans.

But is it as we know it? Undoubtedly not, because what we know is constructed by our brains. This applies both to the material world and to the world of mathematics. So both are different in reality, but there is undoubtedly a connection. But its nature will be hidden to us for all times. Therefore, physicists and mathematicians go the pragmatic way and ignore the fundamentally unknown differences. But I think one should always keep in mind that what we call the material world and the world of mathematics is "colored" by the human brain. We only know the human versions of these worlds. Aliens, if they exist, could therefore have a completely different idea of both worlds.

The mathematical physicist Roger Penrose (born 1931) assumes the existence of three worlds. For him, there is also the world of consciousness. Undoubtedly, the material world is connected to it, for consciousness appears in the brains. But how is it generated? It's all in the dark. Neuroscientists have discovered that there are so-called correlates of consciousness in the brain. This means that a connection can be established between the occurrence of consciousness and the activity of the brain. But where does what we experience through our consciousness come from? For example, the red of a sunset. Neuroscientists have nothing to say about this. They are increasingly able to decipher the structure and function of the brain using ever better investigation methods. But they do not come closer to consciousness itself. Obviously, there is an insurmountable very deep gap between it and what can be uncovered by the methods of natural sciences. This clearly shows that the current scientific view of the world is incomplete. But how should it be supplemented? In the last two chapters, I present a proposal on this. And it provides the key to consciousness.

Heisenberg's uncertainty principle

A famous principle that almost everyone has heard of at least once in their lifetime is **Heisenberg's uncertainty** (or **indetermination) principle**.

Usually it is explained with a few sentences or with wrong analogies and people (sometimes even professional physicists) have a misled understanding of what it is about. Therefore, to get a clear picture and avoid misconceptions, it is instructive to further deepen diffraction and interference phenomena for the one, two, and many slits cases.

The question at this point is: What happens if we use only one slit? Or vice versa: What happens when we use many slits?

By the end of the seventeenth century and well before the birth of QM, the single and N-slits interference phenomena were well known and extensively studied

Chapter 2 : Relation between waves and particles

Apart from the fact that similar statements are already self-contradictory (if a particle **is** a wave, it is not a particle in the first place, eventually one should speaks instead of a particle that acts like a wave—that's why some physicists would like to eliminate any reference to 'dualities' from textbooks altogether), it is high time to debunk the 'observer myth' in QP. No quantum phenomenon needs a human mind or consciousness observing it to validate its existence, and there is no 'particle transmutation' in the sense that our naive intuition tends to suggest.

It is no coincidence that we are presenting wave-particle duality to the reader after dwelling first on a historical and conceptual introduction in order to furnish the context with which to grasp the deeper meaning and implications of the theory. We do the same for Heisenberg's uncertainty principle in the next section.

We have seen that photons can behave either as waves or as particles, according to the context and the experimental arrangement. If the pinhole is small, while it will determine the position of a particle with higher accuracy, it will also produce a relatively broad diffraction **pattern** that renders the position of the photon on the screen uncertain. If the pinhole is small, while it will determine the position of a particle with higher accuracy, it will also produce a relatively broad diffraction **pattern** that renders the position of the photon on the screen uncertain. We can know in a relatively precise manner where the photon or matter particle went through the piece of paper, inside the space constrained by the pinhole's aperture. But it will be displaced laterally on the screen anyway due to diffraction and interference phenomena.

We can know in a relatively precise manner where the photon or matter particle went through the piece of paper, inside the space constrained by the pinhole's aperture. But it will be displaced laterally on the screen anyway due to diffraction and interference phenomena.

For instance, the photoelectric effect or the Compton effect gives one answer, and the double slit experiment and Bragg diffraction give the opposite one. In the latter case we have also seen how the wave character fits with material particles too, like electrons or neutrons.

However, we might ask at this point: When, how, and under which circumstances does the one or the other aspect arise? Is this a property that quantum objects acquire switching from one state to another? Is there somehow a moment in time where a somehow ill-defined entity is or behaves like a particle and afterwards morphs into a wave? If so, does this happen in a continuous fashion, or instantly? Or is there another possible interpretation and understanding which unites the two aspects of photons, as waves and as matter and energy?

In some sense we have already seen how, with the de Broglie hypothesis, it is possible to unify this disparity with the concept of the wave packet.

Therefore, so far we have shed some light (pun intended) on this wave-particle duality, but the picture is still somewhat incomplete. Because, if we think about it carefully, nobody has observed a light wave in the way we can observe a water wave on the surface of an ocean. What we always observe are the effects of something that we imagine, by deduction after the fact, to be something that acts like a wave. The several experiments that hinted at the particle- or wave-character of light did **not** show that light **is** a particle or **is** a wave, but that it **behaves** like a particle or a wave. We should not confuse our human mental (sometimes even ideological) projections of how reality appears, with reality itself.

It is here where Young's famous double slit experiment comes into play again in its modern quantum version. It clarifies further some very important aspects and clarifies the nature of physical objects in QP.

So, let us turn back to the transverse plane wavefront of light, say, just a beam of light coming from a source which ideally is highly monochromatic, that is, having a single or very narrow range of wavelengths and colors (nowadays an easy task to accomplish with laser light).

We know that the slits should have a size and separation which should be comparable with that of the wavelength of the incoming light (something which, even for small wavelengths, can be achieved with high precision using modern lithographic technologies). When the plane wave goes through these slits it is diffracted and is detected with a screen at some distance, and the usual resulting interference pattern appears. These fringes can be easily photographed with a photographic plate or a CCD camera or whatever kind of sensor sensitive to the wavelength we are working with.

Through Young's experiment, the question about the nature of light seemed to receive a clear answer. We conceive of light as a wave since it behaves like a wave, due to displaying the interference phenomenon typical to waves.

However, upon closer inspection at the microscopic level, this turns out to be just an upside-down view of the larger story. How so? Because what we detect are always and only interactions between the light and the detector screen in tiny localized spots. What one observes locally is not some nice continuous shading of the intensity between the black and white fringes, but instead a random distribution of points, that is, of point-like localized interactions on the detector screen, just like pixels on a monitor. For instance, if we use an old conventional photographic plate as our detector screen, develop it, and then look at the photograph under a microscope, we see many tiny white spots which correspond to the grains that, due to a chemical reaction, become white if a photon is absorbed.

After all, when we make a measurement, what we ultimately observe are particles, not waves. Be it a photographic plate, a modern CCD-sensor, or any other kind of detector which responds to an electric signal or just 'clicks'

(like a photodiode or a photomultiplier), what we really measure are one or many dots, local point-like interactions which form a granular interference pattern, never a continuous pattern. Even if we were to build a detector with the highest resolution possible—say, every pixel is an atom—the same dotted interference pattern would emerge at smaller scales. This should not come as a surprise, since from what we have already learned about the blackbody radiation theory, we know that matter, and therefore atoms and molecules, absorb 'quanta' not in a continuous fashion but rather in the form of discrete amounts of energy.

Therefore, also in Young's experiment, we recover and find again the corpuscular nature of light: What seems to be diffracted at the slits and interferes on the screen behaves like a wave, but what is finally observed are photons hitting the screen. On one side, we have something we imagine to be a wave that goes through the double slits; but when we attempt to detect it, it inevitably shows up as a localized interaction.

In a certain sense we find ourselves again at the starting point. What is a photon **really**? If it is a wave, why, when, and how does it become a particle in order to be detected as a point-like structure?

Let us further inspect this state of affairs. If we would strictly maintain the particle picture, then we must assume that each particle must go through one or the other slit, and then travel further in direction of the detector screen and show up at one or the other fringe, that is, in two sets of piled-up particles detected only in two locations on the screen.

However, we know now that this is not the case, since we observe other lines appearing at the interference fringes as well. This naive picture is simply disproved by facts.

Left: What would be expected but is not observed.

Right: What is observed and cannot be reconciled with a classical 'particle-picture'.

One has to assume the existence of some ill-defined 'wave' or 'force' or

'field' that 'guides' the particles along its path, so that they match the interference fringes. Indeed, this is an interpretation that is taken seriously by some physicists, with the so-called **'pilot wave theory'** (described further in our second volume), originally formulated by the American physicist David Bohm (1917-1992). But I strongly suggest you carry on reading and learn what so far is held as true, before jumping into the plethora of interpretations of QM; many of these speculations can boast no hard supporting evidence.

In fact, let us take a further step in our reasoning and ask ourselves what happens if only one photon is sent towards the double slits. Do interference fringes arise also with a single photon? Or are they the result of the collective interaction of many photons? Or, which seems to be another possible alternative interpretation, does the single particle interfere with itself?

The latter question may seem at first meaningless, because we might object that a single photon could not be subjected to interference, since to observe these interference phenomena we need a collection of photons to build up as fringes on the screen. However, there is a way to test this hypothesis: We could try to deceive Nature by seeing whether and how the interference pattern builds up if we send only one photon after another and then wait to see how things develop over time. One after another means that we shoot a single photon at the two slits, wait until the single photon hits the screen and record its position, and only after that, we send the next photon, wait again until it is detected on the screen, and so on.

Chapter 3 : Max Planck the Father of Quantum theory

All objects emit electromagnetic radiation, which is called heat radiation. But we only see them when the objects are very hot. Because then they also emit visible light. Like glowing iron or our sun. Of course, physicists were looking for a formula that would correctly describe the emission of electromagnetic radiation. But it just didn't work out. Then, in 1900, the German physicist Max Planck (1858 - 1947) took a courageous step.

The emission of electromagnetic radiation means the emission of energy. According to the Maxwell equations, this energy release should take place continuously. "Continuously" means that any value is possible for the energy output. Max Planck now assumed that the energy output could only take place in multiples of energy packets, i.e. in steps. That led him to the correct formula. To the energy packets, Planck said "quanta." Therefore the year 1900 is regarded as the year of birth of quantum theory.

Important: Only the emission (and also the absorption) of the electromagnetic radiation should take place in the form of quanta. Planck didn't assume that it itself was composed of quanta. Because that would mean that it would have a particle character. However, like all other physicists of his time, Planck was completely convinced that electromagnetic radiation consisted exclusively of waves. Young's double-slit experiment has shown it and the Maxwell equations have confirmed it.

In 1905 an outsider named Albert Einstein was much more courageous. He took a closer look at the photoelectric effect. It means that electrons can be knocked out of metals by irradiation with light. According to classical physics, the energy of the electrons knocked out should depend on the intensity of the light. Strangely enough, this is not the case. The energy of the electrons does not depend on the intensity, but on the frequency of the light. Einstein could explain that. For this back again to the quanta of Max Planck. The energy of each quantum depends on the frequency of the electromagnetic radiation. The higher the frequency, the greater the energy of the quantum. Einstein now assumed, in contrast to Planck, that the electromagnetic radiation itself consists of quanta. It is the interaction of a single quantum with a single electron on the metal surface that causes this electron to be knocked out. The quantum releases its energy to the electron. Therefore, the energy of the electrons knocked out depends on the frequency of the incident light.

However, the skepticism was great at first. Because electromagnetic radiation would then have both a wave and a particle character. But later another experiment also showed its particle character. This was the experiment with X-rays and electrons carried out by the American physicist Arthur Compton (1892 - 1962) in 1923. As already mentioned, X-rays are also electromagnetic radiation, but they have a much higher frequency than visible light. Therefore the quanta of X-rays are very energetic. That's why they can invade the human body. But that makes them so dangerous. Compton was able to show that X-rays and electrons behave similar to billiard balls when they meet. Which again showed the particle character of the electromagnetic radiation. So their dual nature, the so-called "wave-particle dualism", was finally accepted. By the way, it was Compton who introduced the term "photons" for the quanta of electromagnetic radiation.

What are photons? That is still unclear today. Under no circumstances should they be imagined as small spheres moving forward at the speed of light. Because the photons are not located in space, so they are never at a certain place. Here is a citation from Albert Einstein. Although it dates back to 1951, it also applies to today's situation: "Fifty years of hard thinking have not brought me any closer to the answer to the question "What are light quanta? Today, every Tom, Dick and Harry are imagining they know. But they're wrong."

The Bohr atomic model

We take atoms for granted. In fact, their existence was still controversial until the beginning of the 20th century. But already in the 5th century BC the ancient Greeks, especially Leukipp and his pupil Democritus, spoke of atoms. They thought matter was made up of tiny, indivisible units. They called these atoms (ancient Greek "átomos" = indivisible). In his miracle year 1905 Albert Einstein not only presented the special theory of relativity and solved the mystery of the photoelectric effect, he was also able to explain the Brownian motion. In 1827 the Scottish botanist and physician Robert Brown (1773 - 1858) discovered that dust particles only visible under the microscope make jerky movements in water. Einstein was able to explain this by the fact that much smaller particles, which are not visible even under the microscope, collide in huge numbers with the dust particles and that this is subject to random fluctuations. The latter leads to the jerky movements. The invisible particles must be molecules. The explanation of the Brownian movement was therefore regarded as their validation and thus also as the validation of the atoms.

In 1897, the British physicist Joseph John Thomson (1856 - 1940) discovered electrons as a component of atoms and developed the first atomic model, the so-called raisin cake model. The atoms therefore consist of an evenly distributed positively charged mass in which the negatively charged electrons are embedded like raisins in a cake batter. This was falsified in 1910 by the New Zealand physicist Ernest Rutherford (1871 - 1937). With his experiments at the University of Manchester, he was able to show that the atoms are almost empty. They consist of a tiny positively charged nucleus. Around him are the electrons. They should revolve around the nucleus like the planets revolve around the sun. Another form of movement was inconceivable at that time. That led physics into a deep crisis. Because the electrons have an electrical charge and a circular motion causes them to release energy in the form of electromagnetic radiation. Therefore the electrons should fall into the nucleus. Hence the deep crisis, because there should be no atoms at all.

In 1913 a young colleague of Ernest Rutherford, the Danish physicist Niels Bohr (1885 - 1962), tried to explain the stability of atoms. He transferred the idea of quanta to the orbits of electrons in atoms. This means that there are no arbitrary orbits around the nucleus for the electrons, but that only certain orbits are allowed. Each has a certain energy. Bohr assumed that these permitted orbits were stable because the electrons on them do not emit electromagnetic radiation. Without, however, being able to explain why this should be the case. Nevertheless, his atomic model was initially quite successful because it could explain the so-called Balmer formula. It has been known for some time that atoms only absorb light at certain frequencies. They are called spectral lines. In 1885, the Swiss mathematician and physicist Johann Jakob Balmer (1825 - 1898) found a formula with which the frequencies of the spectral lines could be described correctly. But he couldn't explain them. Bohr then succeeded with his atomic model, at least for the hydrogen atom. This is because electrons can be excited by photons, which causes them to jump on orbits with higher energy. This is the famous quantum leap, the smallest possible leap ever. Since only certain orbits are allowed in the Bohr atomic model, the energy and thus the frequency of the exciting photons must correspond exactly to the energy difference between the initial orbit and the excited orbit. This explained the Balmer formula. But Bohr's atomic model quickly reached its limits because it only worked for the hydrogen atom. The German physicist Arnold Sommerfeld (1868 - 1951) expanded it, but it still represented a rather unconvincing mixture of classical physics and quantum aspects. Besides, it still could not explain why certain orbits of the electrons should be stable.

Sommerfeld had a young assistant, Werner Heisenberg (1901 - 1976), who in his doctoral thesis dealt with the

Bohr atom model extended by Sommerfeld. Of course, he wanted to improve it. In 1924 Heisenberg became assistant to Max Born (1882 - 1970) in Göttingen. The breakthrough came a short time later, in 1925 on the island of Helgoland, where he cured his hay fever. He was able to explain the frequencies of the spectral lines including their intensities using so-called matrices. He published his theory in 1925 together with his boss Max Born and Pascual Jordan (1902 - 1980). This is considered to be the first quantum theory and is called matrix mechanics. I will not explain it in more detail because it's not very clear. And because there is an alternative mathematically equivalent to it. It enjoys much greater acceptance because it is easier to handle. It is called wave mechanics and was developed in 1926, just one year after matrix mechanics, by the Austrian physicist Erwin Schrödinger (1887 - 1961).

The Schrödinger equation

Before we come to the wave mechanics of Erwin Schrödinger, we have to talk about the French physicist Louis de Broglie (1892 - 1987). In his doctoral thesis, which he completed in 1924, he made a bold proposal.

At that time, as explained in the penultimate section, wave-particle dualism was a characteristic exclusively of electromagnetic radiation. Why, so de Broglie, shouldn't it also apply to matter? So why should matter not have also a wave character in addition to its indisputable particle character? The examination board at the famous Sorbonne University in Paris was unsure whether it could approve it and asked Einstein. He was deeply impressed, so that de Broglie got his doctorate. However, he could not present any elaborated theory for the matter waves.

Erwin Schrödinger then succeeded. In 1926 he introduced the equation named after him. The circumstances surrounding its discovery are unusual. It is said that Schrödinger has discovered it at the end of 1925 in Arosa, where he was with his lover.

The Schrödinger equation is at the center of wave mechanics. As already stated, it is mathematically equivalent to Heisenberg's matrix mechanics. But it is preferred because it is much more user-friendly. There is a third version, more abstract, developed by the English physicist Paul Dirac. All three versions together form the non-relativistic quantum theory called quantum mechanics. As you rightly suspect, there is also a relativistic version. I will deal with it briefly in the next section. Only briefly because it brings no new aspect to the topics discussed in this book.

The Schrödinger equation is not an ordinary wave equation, as it is used, for example, to describe water or sound waves. But mathematically it is very similar to a "real" wave equation. Schrödinger could not explain why it is not identical. He had developed it more out of intuition. According to the motto: What could a wave equation for electrons look like? This can also be called creativity. In fact, very often in the history of quantum theory, there was no rigorous derivation. It was more of a trial and error until the equations that produced the desired result were found. It is very strange that a theory of such precision could emerge from this. However, as I will explain in detail, the theory is also mastered by problems that have not yet been solved.

The solutions of the Schrödinger equation are the so-called wave functions. It was only with them that the stability of the atoms could be convincingly explained. Let us consider the simplest atom, the hydrogen atom. It consists of a proton as the nucleus and an electron moving around it. If one solves the Schrödinger equation for the hydrogen atom, then one finds that the electron assumes only specific energy values. It is important that there is a smallest energy value. This means that the electron always has a certain distance from the nucleus. Which makes the hydrogen atom stable. Bohr's atomic model also provides the specific energy values, but cannot justify them. Does that make the Schrödinger equation better? Not really, it is the used mathematical formalism that led to the specific energy values. Why it is exactly this formalism that fits reality, nobody can answer until today.

An object is assigned a state at any time. In classical physics, it consists of an unambiguous position and an unambiguous velocity. This is called determinism, one of the central pillars of classical physics. In quantum mechanics the states are completely different, they are represented by the wave functions. They are therefore abstract quantities from the world of mathematics and not directly observable. But they provide the positions and velocities. And at any time very many different positions and velocities. Therefore, the states are not unambiguous, they are so-called superposition states. As a consequence, quantum mechanics is no longer deterministic. Although that is not entirely uncontroversial, as we will see.

The quantum field theories

The development of quantum theory was not finished with quantum mechanics. Because it has its limits: It doesn't contain the special theory of relativity. Therefore, it only applies to objects that move much slower than the speed of light. What is true for the electrons in the atoms and molecules, therefore quantum mechanics fits there wonderfully. But especially the photons, which always move at the speed of light, are not covered by the Schrödinger equation at all. As already mentioned, the English physicist Paul Dirac has not only developed the abstract version of quantum mechanics. In 1928 he succeeded in integrating the special theory of relativity into the Schrödinger equation. This is the Dirac equation, which I have given as an impressive example (keyword antiparticles) that we discover mathematics and thus the laws of physics and not invent them.

But even with the Dirac equation, the photons could not be described. Moreover, like the Schrödinger equation, it only applies to a constant number of particles. The special theory of relativity, however, makes its creation and destruction possible. Both the photons and the variable particle number required a completely new concept, that's quantum field theory (QFT). The step from quantum mechanics to QFT, however, was only a small step in comparison to the enormous revolution from classical physics to quantum mechanics. And most importantly, QFT cannot help solve the problems I will discuss in the next chapter.

Their development began already in the 20's, parallel to quantum mechanics. Paul Dirac, Werner Heisenberg and Pascual Jordan were significantly involved. You already know them. Also the Italian physicist Enrico Fermi (1901 - 1954) and the Austrian physicist Wolfgang Pauli (1900 - 1958) made contributions. But the development quickly came to a standstill because there were absurdities in the form of infinitely large intermediate values that could not be eliminated for a long time. It was not until 1946 that people learned to deal with them. The first quantum field theory emerged around 1950, it was quantum electrodynamics (QED), the quantum version of the Maxwell equations. The American physicist Richard Feynman (1918 - 1988), a charismatic personality, played a decisive role in its development. He tended to bizarre actions, for example drummed regularly in a nightclub. And he was involved in solving the Challenger disaster in 1986. Richard

Feynman is the only physicist who gave still lectures for beginners when he was already famous. That was in the early '60s. They have also been published in book form and are still widely used today. The German textbooks on physics are, as expected, factual and dry. The Feynman textbooks are more „relaxed" and contain much more text. My impression: For the first contact with physics they are less suitable, but they improve its understanding, if one has already learned some physics.

QED is the foundation of everything that surrounds us. The entire chemistry and thus also biology follows from it. But even with it, the development of quantum theory was not yet complete. Because two more forces were discovered, the strong and the weak force. They only play a role in the atomic nuclei, so we don't notice them. We only notice the gravitational and the electromagnetic force. Even though large bodies are always electrically neutral, QED also plays an important role in everyday life. Matter is almost empty. Accordingly, if two vehicles collide, they should penetrate each other. If it wasn't for QED.

The strong and the weak force led to the development of two further QFTs. Whereby the QFT of the weak force could be combined with the QED. For the sake of completeness, the QFT of the strong force is called quantum chromodynamics (QCD). The QFTs of the three forces that can be described with them are combined to form the so-called standard model of particle physics. It represents the basis of modern physics.

In the QFTs, each type of elementary particle is described as a field in which particles = quanta are created and destroyed. This is the core of the QFTs. What are the elementary particles? Matter consists of molecules that consist of atoms that can be separated into nucleus and electrons. The electrons are called elementary particles because they cannot be further divided according to today's knowledge. The nucleus is composed of protons and neutrons, each consisting of three quarks, or more precisely of different mixtures of up and down quarks. They are elementary particles. Ordinary matter thus consists of three different elementary particles. But there are many more matter elementary particles, 24 in total, and that's not all either, because three of the four different forces also consist of elementary particles. Only the gravitational force cannot be integrated into the concept of the QFTs.

And there is another elementary particle, the Higgs particle, which was discovered at the Large Hadron Collider (LHC) in Geneva in 2012. It plays a special role, I won't go into it further **[10]**.

Since the general theory of relativity cannot be integrated into the standard model of particle physics, there must be a more fundamental theory. String theory was a promising candidate, but hope is fading **[11]**.

Will the more fundamental theory, if it is ever found, be the "theory of everything"? So will it herald the end of theoretical physics? Some believe this, such as the recently deceased physicist Stephen Hawking (1942 - 2018). So far, however, every physical theory always has something inexplicable in it. One can hope that we will find a theory that explains everything. But I doubt that we will ever succeed. Therefore, even in the far future, there should be theoretical physicists who are looking for new theories.

Chapter 4 : Laws Of Quantum Physics

As every good textbook on electromagnetism will tell you, a single slit, or a pinhole, will produce an interference pattern, though a less pronounced one than a larger slit produces. It is an easy experiment that you can do by yourself—just take a piece of paper, poke a little round hole with a pin or needle, and look towards a light source. You will observe the image of the light source surrounded by several concentric colored fringes (the colors appear only due to the fact that, fortunately, the world we live in is not monochromatic).

Due to the diffraction that occurs whenever a wave encounters an object or a slit, especially when its size is comparable to the wavelength, the plane wave front is converted to a spherical or a distorted wave front, which then travels towards the detector screen.

In this case you see that, even though we are dealing with only one slit, some weak but still clearly visible secondary minima and maxima fringes can be observed.

An important point to keep in mind is to avoid a common misconception (which is frequently promulgated in some popular science books) which states that only two or more slits can produce interference fringes, whereas, for the single slit, interference effects disappear. This is not entirely correct.

True, it is easier to produce more pronounced interference patterns with more than one slit (or pinhole); and for most applications, especially when the wavelength of the incident wave is much smaller than the size of the aperture, these effects can be neglected.

Transverse plane wave incident on a single slit with aperture a=5□.

However, strictly speaking, a single slit also produces small diffraction and interference phenomena.

An elegant explanation of how interference comes into being, also for a single slit, dates back to the French physicist A. J. Fresnel. He borrowed an idea from Huygens (hence the name ` Huygens Fresnel **principle**'), according to which every single point on a wave front should itself be considered a point-source of a spherical wave.

Along the slit's aperture emit at the same time their own spherical wave fronts which, however, when seen from a position on the screen, add up to produce an interference pattern. The reason for this is not so difficult to visualize.

Since all fronts are initiated in different locations along the aperture they will also travel a different path length, which implies that they have different phase shifts when they overlap on the screen.

For instance, where we saw the two paths of the two sources from the edges. As in the case of the double slit, these two rays have a relative phase shift by an amount Δ and, when superimposed together on the screen, they form a resultant intensity according to the interference laws. In this case, however, this holds not only for two waves, but for an infinite number of point-sources along the aperture. Fresnel, by making the appropriate calculations, was able to show that if one sums up all of the spherical wave fronts coming from the points of the aperture of the single slit and projects these onto all the points along the detector screen, then one obtains indeed the known diffraction and interference patterns.

The Huygens-Fresnel principle

If we repeat the same experiment with a slit that has a size close to the wavelength of our incoming wave front, then we see that the interference fringes disappear.

Only when the size of the slit is equal to or smaller than the wavelength are the fringes definitely absent. This is because the slit is so small that only a single point-source can form a spherical wave front with a wavelength equal to the slit size, and there can be no path difference and phase shift with some other source which could produce the interference pattern. However, diffraction has become very large instead, so that the photons will displace themselves on a relatively large area on the detector screen, according to a bell-shaped distribution called the 'diffraction **envelope**'.

The parameters that determine the angular dependence of the interference pattern are: first, the size of the aperture relative to the wavelength (here: **a**=3☐); second, the spacing **d** between the slits (here: **d**=3 **a**); and, of course, the number of slits. The three curves represent respectively the 1, 2, and 10 slits diffraction cases. The intensities have been normalized in all the cases to unity.

For the one slit case you see there are some weak but clearly discernible secondary lateral peaks. They reduce almost to the diffraction envelope.

For the two slits, as in the case of Young's double slit experiment, we obtain more pronounced fringes. You can see how the one slit pattern 'envelops' the two slits pattern. The interference figure of the single slit pattern is split into several more fringes. However, notice that it would be incorrect to say, as you might hear frequently, that when we switch from the double slit to the single slit case, interference phenomena disappear. That is, in general, not the case. What happens is that we return to the envelope of the single slit which contains many fewer fringes, but still might have some other interference fringes too (and in this case it does). Again, interference is not a phenomenon specific to the double (or more) slits experiment.

Interference does not disappear if one slit is covered; it merely becomes weaker than it is with more slits.

Finally, in the case of 10 slits, the two slits curve turns out to be the envelope of the 10 slits curve. So, you can observe how this is a more general trend and phenomenon which results from the interplay between diffraction and interference. In fact, generally, the N-slits fringes and their spacing's arise due to this combined effect between diffraction and interference.

These were only a few examples to outline, at least intuitively, how wave interference works. Programmers who might be interested in recreating the infinite possible combinations of interference figures can find the general formula of the intensity function in Appendix a VI.

We have seen in this section, especially with Bragg diffraction, how diffraction and interference phenomena determine the wave-particle duality not only of light particles, photons, but more generally of all particles—including material particles that have a mass and that behave like waves, even though we think of them as localized pieces of solid matter.

A question we might ask is: What happens with a particle if we want to know its precise whereabouts in space? For example, let us determine the precise position of a particle by letting it go through a tiny pinhole, as Isaac Newton did with photons in his investigations of the nature of light. If a particle goes through that single little hole on a piece of paper, we are authorized to say that we can determine its precise position in space. In fact, this can be done, but at a cost. Because, on the other hand, we know that due to the wave-particle duality, we can't forget the particle's wavy aspect. When a particle, also a material particle, goes through this pinhole, it will likewise be diffracted and afterward position itself on the detector screen according to an interference pattern.

If, instead of dealing with slits, we take a tiny round hole of a size comparable to that of a few multiples of the wavelengths, we obtain circular interference fringes. This is an intrinsic and unavoidable effect for all types of waves.

The pinhole as a detector of a particle's position can't avoid interference.

If particles must be linked to a wave, according to the de Broglie relation, conceiving of it as a wave packet, we will always have the interference fringes, even with only one hole or slit and even with only one particle.

Compares the two cases in which the pinhole determines the position of the particle with an aperture of a=2λ(high precision) and a=20λ(low precision), with □, as usual, the wavelength of the photon or, in case of a matter wave, the de Broglie wavelength.

If the pinhole is small, while it will determine the position of a particle with higher accuracy, it will also produce a relatively broad diffraction **pattern** that renders the position of the photon on the screen uncertain. We can know in a relatively precise manner where the photon or matter particle went through the piece of paper, inside the space constrained by the pinhole's aperture. But it will be displaced laterally on the screen anyway due to diffraction and interference phenomena.

Of course, with a single particle producing a single spot on the screen, no interference figure is visible. However, as we have learned with the single photon diffraction at the double slit, the probability of finding this spot is in one-to-one correspondence with the intensity of the interference fringes that many particles produce.

Moreover, recall also that we can't predict where precisely this spot will appear.

If, instead, the pinhole is large, fringes will become less pronounced. We will know where the photon will hit the screen with relatively good accuracy, which means it 'felt' only a tiny displacement along the screen.

However, by doing so, we will lose our capacity to determine where exactly the particle went through the pinhole, as it is no longer a pinhole at all, but a large hole. There is no way, never ever, not even in principle, to obtain the precise measurement of the particle position and at the same time avoid the production of interference fringes (or interference circles, as in the case of a circular aperture or large diffraction effects). One will always obtain a more or less pronounced bell-shaped or peaked distribution of white spots on the screen. This is not because we don't have a sufficiently precise measurement apparatus but because it is a consequence of the intrinsic wave-nature of particles. It is a universal law of Nature, according to which it is hopeless to believe that we can pass a wave through a slit and not observe any interference and diffraction phenomena.

Heisenberg's uncertainty principle explained using the single silt (or pinhole) diffraction experiment.

Now, this could also be interpreted as follows: The particle, once it has gone through the slit, will acquire an extra momentum, λp, along the vertical axis. This does not happen because of the interaction of an outside force or, as we might imagine naively, by an interaction, deviation, or bouncing effect of the particle with the slit's edges because, in that case, we would observe a random distribution but not an interference pattern. This extra momentum

λp which displaces the particle along the detection screen is due only and exclusively to the wave nature of matter and light. We might interpret this also as a 'scattering' of the particle but we should keep in mind that this is misleading terminology, as there is no scattering force at all. No scattering interaction or forces from the outside are necessary to make this happen.

Where does this extra amount of momentum λp come from? It is simply the uncertainty we have about the particle's momentum in the first place. It is an inherent uncertainty of the properties of any particle due to their wave nature. This is the only possible conclusion if we want to avoid violating the principles of the conservation of momentum and energy.

The point is, we will never be able to determine with extreme precision

– That is, with an infinitely small slit of size λx=0 – without blurring the momentum because, by doing so, we will inevitably diffract the plane wave front, the wavelength of which is given by the de Broglie's relation.

This will inevitably displace it according to a statistical law which reflects the diffraction and interference laws.

So, we must conclude that the smaller our uncertainty in determining the particle position (the size λx of the slit), the larger the diffraction effects and, therefore, the larger the uncertainty over the momentum. (λp becomes large in the vertical direction.) On the other hand, if I want to know the particle's momentum with small uncertainty (λp small), we will have to open up the slit's aperture (λx large) to reduce the diffraction. However, I will never be able to determine with precision both the momentum and the position of a particle at the same time. We have to choose whether we want to keep focused on one or the other; never ever are we allowed obtaining both. Again, this is not because we are perturbing the system but because we are dealing with waves.

We have, however, interference fringes here, and in principle, the particle can displace itself very far from the center of the screen, especially if the slit becomes very small. However, this also happens with a decreasing probability. It is much more probable to find the particle hitting the screen at the central white peaks than at the little ones far from the center. For practical reasons, one must set a limit as a convention. Due to statistical reasons (which will be clarified in chapter III 3), one takes as convention the ' **normal distribution**' which states that λp is the width of the central fringe corresponding to 68.3% of all the particles it contains. That is because this is the central region with higher intensity and, therefore, that region where we have the highest probability of detecting a single particle. In fact, as we have seen with the single slit, the secondary interference fringes beyond the central one are usually quite small.

So, this means that, to a specific amount of λx, we must expect to find also a specific amount of λp – always and inevitably, by the laws of Nature.

Heisenberg was able to show that the product of these two uncertain quantities is a constant and he summed this up with his very famous and extremely important inequality, which states:

$\lambda \cdot \lambda \geq \hbar/2$ **Eq. 8**

Where $\hbar =$ is Planck's constant divided by 2п, called the **'reduced Planck constant'** (read "h-bar"). This is a mathematical way of stating what we have said so far: If you force the value of the position of a particle to be tight, it will always have an uncertainty over the momentum of $\lambda p \geq \hbar$,

$\cdot \lambda$

That is, a precise measurement of the particle position wills simply an uncertainty over its momentum. (A small λx induces λp to become large.) Vice versa, if you force the value of the momentum of a particle into a narrow range of values, it will always have an uncertainty over the position of $\lambda x \geq \hbar$.

·λ

This is an inequality which can be obtained in several ways. One closely related to this is via the Fourier transforms, that is, a mathematical tool developed by French mathematician Joseph Fourier (1768 –1830) who showed how every time series (a signal in time, such as a complicated traveling wave) can be decomposed into the sine and cosine functions of different frequencies. In modern QT, Heisenberg's inequality is obtained through a mathematical operator theory approach, which is mathematical y more rigorous but somehow obscures the deeper meaning of Heisenberg's principle related to the wave nature of matter. At any rate, always keep in mind the little but very important inequality of Eq. 8, as it will turn up more or less everywhere in QM.

Now let us analyze what interpretation Heisenberg gave to his own principle. He illustrated this with a famous 'thought experiment'. By 'thought experiment' (from the German 'Gedanken experiment'), in science one means an ideal experiment which could be realized in principle and does not violate the laws of physics but can't be performed in practice due to technological limitations or other constraints. Heisenberg's thought experiment elucidates what nowadays is remembered as **'Heisenberg's microscope interpretation'**. It is an interpretation that, unfortunately, several professional physicists have also adopted (probably to make it more understandable to the public or, being busy only with calculations, to avoid digressions in the conceptual foundations that might trigger annoying questions from too curious students). However, if you have carefully followed what we have said so far, you will be able to recognize that Heisenberg's interpretation is somewhat misleading. It went as follows.

Imagine that we want to detect the position and momentum of a particle in a region x by using an incoming photon which hits the particle. This photon is then scattered toward some direction.

The old Heisenberg's microscope interpretation.

We have already described something similar with the Compton scattering effect. Then, as was Heisenberg's reasoning, we can look to where the photon has been scattered to deduce the electron's position

However, according to classical optics, the larger the angular aperture θ of the microscope, the better it will resolve the position of the scattered photon and, therefore, deduce the position of the electron. Yet this angular aperture depends not only on the size of the microscope lens but also on the wavelength of the photon. To have high precision in determining from which direction the photon comes, that is, in determining the position of the electron, we must use a photon with a short wavelength. However, if the photon has a short wavelength (high frequency), it will carry more momentum because it is more energetic due to Planck's equation.

Therefore, the kick that it imparts to the electron will also be larger. That is, if I want to observe the precise position of the electron, I must use photons which will produce a large recoil on the electron itself. This makes its position again indeterminate. On the other hand, if I want to more precisely determine the momentum of a particle, I'm forced to use photons with large wavelengths because only they have a small momentum and will not displace the electron too much through Compton scattering. Yet if I use long wavelength photons, the microscope optics will resolve their position with less accuracy. So, again, we must search for a tradeoff between the wavelength we use (the precision over the momentum p) and the microscope aperture (the ability to look at a precise position x).

In the macroscopic world, this effect does not play any role for practical purposes because objects are huge and not disturbed by tiny photons (even though, strictly speaking, it always holds). However, for atoms and even more for particles like electrons, a single photon can disturb them and scatter them away, preventing us from determining their properties. That's why Heisenberg's uncertainty principle is invoked to explain that in the microscopic world, a measurement always perturbs a system in such a way that the scale of the measurement's perturbation is about the same magnitude of the effect it produces and, therefore, renders the result imprecise if not entirely useless.

Heisenberg was able to show that, by reasoning in such a manner, making the appropriate calculations which resort to classical optics, he could indeed re obtain his famous inequality over momentum and position of Eq. 8.

Heisenberg's microscope thought experiment boils down to the following argument: We cannot determine the position and momentum of a particle at the same time because when we make an observation, we must inevitably interact with the object we want to observe (such as sending other particles of comparable mass or energy to the particle being observed). This inevitably disturbs the system, which then loses its original position and momentum that we sought to determine.

This argument is correct insofar that every interaction with microscopic particles inevitably causes them to scatter and prevents us from knowing their precise position and momentum. In spite of that, it becomes wrong when it is invoked as a restatement supposed to explain the origin and cause of quantum uncertainty. Because, that would be in stark contrast to the wave-particle nature of matter it and which doesn't need whatsoever interactions with the outside world to be described consistently. It is an interpretation that does not stand up to the tests of modern QT. The popular belief that Heisenberg's uncertainty principle is about the interaction between the observer and the observed object is wrong. Heisenberg's uncertainty principle is a fundamental law of Nature whose roots are in the wavy nature of matter and has nothing to do with the interaction between the measuring apparatus and the measured objects. Of course, the interactions and imperfections of measurement devices must always be taken into account in a real laboratory but these add further uncertainty to the intrinsic quantum fuzziness that is present a priori. In fact, we shall see later that, as strange as it might sound, it is nevertheless perfectly possible to set up experiments which make measurements without necessarily interacting with the system, and yet the Heisenberg uncertainty principle remains inescapable. It also holds in the case in which we are able to reduce to zero any interaction with the observed object. It is an inherent law of Nature that is independent of observational interactions.

At the time of Heisenberg, it was still legitimate to think of particles in this way. Heisenberg can be excused because several experiments which showed how his own interpretation must be revised came much later – some not even until before the 1990s, with the development of sophisticated laser and quantum optics devices. So, while the historical context and the lack of more experimental evidence justify Heisenberg, it does not do the same with modern physicists. Nowadays, we can no longer stick with the microscope interpretation as a correct understanding of the workings of the uncertainty principle. This is no less fundamental than Bohr's atomic model, the geocentric model or a flat earth theory.

We should accept that Nature is telling us that we should never forget the wave-particle duality. When a particle goes through a pinhole, we must think of it as a physical process described by a transverse planar matter wave, which is diffracted like any other wave and produces a spherical wave front

Forget about classical understandings of particles which possess definite properties such as a position, and which move along precise deterministic trajectories and possess a definite momentum. There are no positions or momenta which describe a particle.

Instead, we must seriously consider these to be emergent qualities, not intrinsic properties. Particles do not at all have a clear and precise position and momentum as we imagine them to have at our macroscopic level and as our naive intuitive understanding wants to make us believe.

QP seems to suggest that the physical objects we imagine as point-like hard material billiard balls are, instead, entities which are intrinsically somewhat fuzzy, with no sharp and well-defined boundaries and that travel throughout space as waves that eventually collapse.

The quantum interpretation of the uncertainty principle.

This nature of particles is independent of the precision of our measurement apparatus and independent of the fact that we interact (or do not interact) by observing it.

Indeterminacy is not a matter of ignorance. Heisenberg's uncertainty principle is not a principle about a lack of information. It is not about particles whose whereabouts and speed we cannot determine. It is about physical entities that simply don't have anything to do with our anthropocentric imagination. This is what distinguishes classical from quantum physics.

Chapter 5 : Quantum Field Theory

Quantum field theory (QFT) integrates special relativity and quantum mechanics, thus providing a description of subatomic particles and their interactions that is more accurate than that provided by quantum mechanics. In QFT particles are viewed as excited states of the underlying physical field: excitations of the Dirac field manifest themselves as electrons; excitations of the electromagnetic field give rise to photons. Moreover, interactions of particles, such as those occurring in high speed collision experiments in an accelerator, are described by perturbative or iterative terms among the corresponding quantum fields. At each level of iterative approximation these interactions can be distilled into a diagram, known as a Feynman diagram; the diagram shows that particle interactions take place via an exchange of mediator or force carrier particles. Quantum electrodynamics (QED) is the first quantum field theory to be developed; QED combines the relativistic theory of electrons (encapsulated in the Dirac equation) and the quantized electromagnetic field in the form of a Lagrangian. The QED Lagrangian with the aid of Feynman calculus leads to scattering amplitude calculations; these calculations represent the most accurate physical predictions in all of physics. Suffice it to say that without QFT we would not have the understanding of particle physics we have today. It is natural to take up QFT once one has gone through quantum mechanics and special relativity theory. At the same time, QFT is considered to be one of the most difficult subjects in modern physics---the theory seems contrived and unnecessarily complex at times, mixing a disparate set of ideas in a series of difficult calculations. This book is a modest but rigorous introduction to QFT: it is modest in its coverage (we work almost exclusively with scalar field theory and QED; moreover, the topic of renormalization is discussed only in passing); it is rigorous in that a detailed mathematical derivation is often provided along with a number of worked out examples. I have tried to provide a concise and comprehensible introduction to QFT without abandoning the ideal of mathematical rigor. Mathematics prerequisites for the book consist of tensor algebra, Fourier analysis, and representation theory; physics prerequisites are a course in quantum mechanics and special relativity. These prerequisites are covered, often without proofs but always with examples, in the first two chapters of the book. Balancing the goal of readability and that of precision is a challenge when writing about modern physics, especially in a limited number of pages.

Chapter 6: Hydrogen Atom

Sommerfeld introduced a second quantum number, the 'orbital' (or

'azimuthal') quantum number l, which for each principal quantum number n specifies n allowed orbital angular momenta of the electron around the atom's nucleus (l=0, 1, 2,...n-1, and designated as s, p, d, f, g,...), and which characterizes the ratio between the semi-major and semi-minor axis—that is, the elongation of the corresponding electron orbit.

This was quite an exciting success at the time, as it seemed to indicate that physicists were after something that could lead them in the right direction. However, many inconsistencies soon became apparent. The Bohr-Sommerfeld model was able to make these predictions with only limited precision and, at any rate, only for the hydrogen spectra. Such a simple theory cannot explain all the other elements of the periodic table, nor can it explain the molecular energy levels and structure.

It was an attempt to explain atomic behavior and energy states. It is, indeed, an interesting historical remnant of the intellectual journey of humankind, and it made perfect sense when viewed in that historical context.

However, it is no longer tenable today. Unfortunately, it is still taught in schools as if it were the final sentence and summation of atomic physics.

And it was a good example which shows that some theories can make correct new predictions and yet still be wrong. This is something that many scientists seem to forget but that we should always keep in mind when working with modern theories.

The Bohr-Sommerfeld hydrogen model ($n=5$). Arnold Sommerfeld (1868-1951).

QM will explain things much more accurately, as we shall see later. It was only upon the advent of QP and its application to the atomic structure that science could correctly predict atomic spectra. The atomic model that arises from QM will be completely different from that of Bohr, which could be considered only a temporary sketch, as Bohr himself had to admit.

4. Even more evidence for particles: the Franck-Hertz experiment, Compton scattering and pair production

Another nice and clear-cut piece of experimental evidence that atoms must have discrete energy levels came only a year later, from the so-called

'Franck-Hertz experiment'. G. L. Hertz was the nephew of H. R. Hertz, the German physicist mentioned as the discoverer of the photoelectric effect.

The experimental setup consists of Hg gas atoms (Hg is the symbol for the element mercury) inside a low-pressure bulb. An electric cathode—that is, something like the heated filament of a light bulb—emits electrons.

Therefore, this part of the device emits not only light but also negatively charged particles. An electric field is applied between the electron emitting cathode and a positively poled grating, with a battery or some other electric source, which builds up an electric potential.

Gustav Ludwig Hertz James Franck (1887-1975). (1882-1964).

Due to their negative charge, this difference in the electric potential field leads to the electrons' acceleration and conveys to them some kinetic energy (as you might recall from school, charges with the same polarity repel each other whereas those with opposite polarity attract each other). When the electrons reach the grating, most of them will fly through because the mesh of the grating is kept sufficiently wide to allow for that. This first part functioned as a little electron accelerator. Then, between the grating and a collecting plate on the right side, another field is applied.

The Franck-Hertz experiment setup of 1914.

However, in this second part of their journey, they will experience an inversely polarized electric field as, after passing the grating, they will be repelled because they will begin to 'feel' the negatively charged collecting plate.

Measures electric current (the number of electrons), one can measure the flux of electrons which flow between the grating and the collecting plate. While the electrons' initial energy is proportional to the applied electric field intensity (the voltage) between the emitting cathode and the grating, in this second part, they are decelerated due to the opposite polarity. The measurement of the current, therefore, allows one to determine the number of electrons that make it through to the collecting plate, which provides information about how their energy is affected by the atoms while flying through the gas in this second part of the bulb. This is done by varying that field, step by step, for several voltages.

Franck and Hertz's insight was that, while flying through the gas of atoms, several electrons must sooner or later hit one or more atoms and be scattered either elastically or inelastically. **'Elastic scattering'** means that when objects hit a target, they change course but maintain the same kinetic energy, while **'inelastic scattering'** implies that they lose part or all of their kinetic energy in the collision process (more on this in Appendix A III). It follows that there must be a measurable difference between the energy of the injected electrons reaching the grating and the energy of those which flew through the gas, hitting the collecting plate. This difference is made clear to the observer by measuring the current between the grating and the collecting plate. This energy gap tells us something about the energy absorbed by the atoms in the gas.

Therefore, if atoms absorb energy only in the form of quanta, this implies that, while we slowly increase the kinetic energy of the injected electrons, we should be able to observe when and to what degree the electrons' energy is absorbed by the gas of Hg atoms.

While the injected electrons' kinetic energy is increased steadily by application of an electric potential from 0 to about 15 V between the cathode and the grating (horizontal axis), the current of the electrons measured at the collecting plate (vertical axis) increases accordingly, though not in a linear fashion. We observe that the electrons do not have a final kinetic energy which increases proportionally to the electrons' input energy, according to what one would expect for an elastic scattering between classical objects (think, for example, of billiard balls). What we see instead is that at first (between 0 and 4 V), the relation between the input and output energy is approximately linear, which means that the electrons are scattered through the gas elastically; they do not lose considerable kinetic energy.

At about 4.5 V, the first bump appears. Between 4 and 5 V, the output energy of the electrons decreases steadily, despite their increasing initial energy. This signals an inelastic scattering: Some of the electrons' initial energy must have suddenly been absorbed in collisions with the Hg atoms.

However, this does not happen before a certain kinetic energy threshold of the electrons hitting the Hg atoms. At about 5.8 V, almost all the kinetic energy is lost and goes into the internal excitation of the atoms. There is, however, a remaining minimum energy gap which is shown in the figure with the vertical arrow. The difference between the first peak and the first minimum is the maximum amount of kinetic energy the atoms are able to absorb from the electrons. Therefore, it furnishes the first excited energy level of the Hg atom. Then, after about 6-9 V, the energy begins to increase again, meaning that the atoms absorb only that aforementioned discrete amount of the electrons' energy, but not more than that. The remaining energy goes again into elastic scattering. All this repeats regularly at about 9-10 V and about 14 V.

The existence of these 'bumps' at different input energies (until nowadays, experimental particle physics is all about the search for bumps appearing in graphs) means that atoms must have several different but discrete energy levels. Franck-Hertz's was the first direct experimental proof confirming Planck's idea that matter absorbs energy in discrete quanta.

Moreover, this validated the discrete spectral lines of light spectra, as did Bohr's idea of representing the atoms in the form of a model which resembles a tiny solar system—that is, with electrons moving only in specific orbits with their respective quantum numbers which represent different but discrete energy levels.

Not too many years later, other types of phenomena confirmed energy quantization in Nature. One of these, in 1923, was the so-called '**Compton scattering**', documented by the American physicist A.H. Compton.

Compton also started from the assumption that EM waves could be considered a flux of light particles. If this were true, it would then be possible to calculate precisely the scattering process between a single photon and an electron, just like it is possible to describe the elastic collisions between two billiard balls, using the simple laws of energy and momentum conservation of CP.

Now, by using these conservation laws, Compton wrote a **Compton** concise and useful formula which relates the **(1892-1962).** Wavelength of an incoming high-energy gamma-ray photon (a photon with a wavelength sufficiently small to be comparable to the size of an electron) before the scattering with the electron, and the wavelength λ' of the scattered photon, according to a scattering angle. The wavelength λ' of the scattered photon must be larger than that of the incoming one, as it loses some of its energy. This is because the higher the energy of a photon, the smaller its wavelength will be. In fact, remember that the energy of a photon is given by Planck's equation $= h\,\pmb{\lambda}$, with $\pmb{\lambda}$ the frequency. The wavelength λ of an EM wave is given by $=$, with $= 3\ 10$ (ca.

300.000 km/s), the speed of light in vacuum. So, the wavelength we associate with a photon must be inversely proportional to its energy as:

$= h\,\lambda = h$. **Eq. 3**

The Compton scattering process.

At this point, the reader might be somewhat confused by the fact that we are talking about light particles, which are supposed to be point-like objects, and yet we continue to treat them (and even depict them graphically) as waves having a specific wavelength. Again, as we have already seen with the photoelectric effect, Nature seems to play with this ambiguity. It seems as if light, or any EM energy traveling throughout space, can be thought of as a wave and, because matter absorbs and emits it in the form of discrete energy quanta, also as a particle. We shall clarify the deeper meaning of this 'duality' in the coming sections. For now, please accept it on faith and follow the line of reasoning and phenomenology of these historical experiments, which became the foundation of all that will be explained later.

By putting all of this together, Compton was able to predict in 1923 a very strict relation between the difference in wavelength of the incoming and scattered photons and the scattering angle , which is the wavelength shift Δ radiation undergoes when it is scattered by matter. Compton's formula was as follows:

Δ = − = (1 − cos).

Here = is a constant and is the mass of the scattering particle, in this case the electron.

Notice that if is zero, we have no difference, which means that the photon goes straight through without scattering and frequency change. Whereas for =180°, for backscattering, the wavelength increases by twice the Compton wavelength. The exact relation between the input and output wavelengths (or energies) of the particles and their respective scattering angle was verified experimentally, and it turned out that Compton's predictions came to fruition precisely. This is a great historic example of the triumph of theoretical physics confirmed by experimental verification—of how, in the history of science, we have found that sometimes first comes the math and then comes the verification in the lab (even though sometimes it goes the other way around).

Compton scattering is also an atomic scattering process. This suggests an explanation of how photons can ionize atoms. To ionize an atom means that a photon extracts an electron from an atom's outer orbit shell around its nucleus. A light particle with sufficiently high frequency can be absorbed by one of the electrons, so that the electron acquires a certain amount of kinetic energy and can eventually be removed by overcoming the atomic force potential that keeps it bound to the nucleus. A gamma-ray photon, however, can be so penetrating that it can ionize even inner electron shells and partial y or completely transfer its EM energy to the electron, which acquires the gamma ray's momentum and kinetic energy and which for this reason is also called a 'photoelectron' (as in the photoelectric effect). Notice however that the photoelectric and the Compton effect are two very different physical processes. The former extracts the electrons from a metal lattice; the latter is a deep scattering process that can 'kick out' the electrons which are contained in the inner shells of an atom.

A single high-energy photon can also be scattered by several atoms and will lose its energy with each repeated collision. A high-energy small-wavelength (high-frequency) photon can become, via multiple scatterings whereby it loses a bit of its energy in each collision, a low-energy large-wavelength (low-frequency) photon.

This is precisely what happens in the center of the Sun. In the Sun's core, which has a temperature of about 16 million degrees, there is a huge amount of gamma radiation. However, it takes several million years of scattering processes before the photons produced inside the Sun reach its surface, by which time they have lost most of their energy being 'downgraded' to the photons that we know as visible light. The light we observe today coming from the Sun's surface was once intense radiation foam of gamma photons in its interior.

Compton scattering of outer atomic shell electrons.

What makes Compton scattering so important and interesting for our considerations here is that all this seems to suggest again that we can reasonably think of EM waves no longer as waves at all, but rather in the context of light particles, or photons, kicking around other particles like tiny billiard balls.

If we were to stick to the classical idea that light is made of waves, we should observe a scattering of concentric waves by the electron (think of water waves scattered by an object). Compton's scattering instead indicates that the application of the conservation laws of CP, which determine the scattering between classical particles, holds also with photons. There is therefore no longer any reason to reject a 'particle picture' in QM in favor of the 'wave picture'.

Another effect which is worth mentioning is the ' **pair creation**' effect, which can occur as an alternative to Compton scattering. Pair creation is a physical phenomenon whereby a high-energy massless photon is converted into particles with a mass. Instead of the photon and electron being scattered, what occurs in this case is the absorption of the photon (by another particle or atom nucleus) and an immediate release of its energy in the form of material particles.

The pair creation effect was discovered in 1933 by the British physicist P. Blackett. This phenomenon clearly shows that we must conceive of light not only as made of photons but also in the context that photons can transform into other material particles, like electrons.

Pair creation, or even **'annihilation'** , which is the opposite process whereby particles with mass are converted into photons, is a striking example of Einstein's **mass-energy equivalence** which is described by his famous formula where **m** is the **rest mass** of a particle, and which is called also the 'rest energy' (in relativity, the mass of a fast-moving object cannot be handled as the classical rest mass, but we won't go further into this here) and, again, **c** is the speed of light. The mass-energy equivalence tells us that every particle, atom, and any material object with a mass contains an energy in potential form which is given by Einstein's formula. The fact that it factors in the speed of light squared makes the amount of energy contained in matter huge.

Only a gram of matter, if converted into kinetic energy, would cause an explosion like that of the bomb of Hiroshima.

This process, however, can only happen if the energy of the incoming photon is sufficiently large to produce the two electrons according to Einstein's mass-energy equivalence formula. If the photon's energy is lower than twice the rest energy of two electrons, the process cannot occur and the photon will not be absorbed but instead eventually will be scattered.

Pair creation (left): a gamma-photon converts into an electron and anti-electron. Pair annihilation (right): an electron and anti-electron convert into two gamma photons.

Fortunately, for some reason that is still not entirely clear, the Universe we live in is made almost completely of one type of particle; otherwise, we would have already 'evaporated away' in a foamy Universe where particles annihilate and create themselves continuously and stars, planets, and of course life, at least as we know it, couldn't have formed in the first place.

Thus far, we have listed several phenomena which indicate how energy is absorbed and emitted in a quantized manner and which therefore led to a corpuscular interpretation of light: Blackbody radiation, the photoelectric effect, the Franck-Hertz experiment, the Compton effect, and pair creation and annihilation can all be viewed in this context.

Chapter 7 : Einstein's Theory Of Relativity

Interestingly, "relativity" was only one of a couple of different names Einstein considered to describe his theory. "Invariance" was another, and he spoke of his "invariance theory" often. If the pinhole is small, while it will determine the position of a particle with higher accuracy, it will also produce a relatively broad diffraction **pattern** that renders the position of the photon on the screen uncertain. We can know in a relatively precise manner where the photon or matter particle went through the piece of paper, inside the space constrained by the pinhole's aperture. But it will be displaced laterally on the screen anyway due to diffraction and interference phenomena.

It was fellow physicist Max Planck, who discovered Planck's constant and together with Einstein helped intuit all the framework of theoretical physics we rely on today, who settled its destiny by referring to the theory as "relativity" in his own publications.

The Maxwell equations contain the speed of light. Normally speeds are always related to something, they are "relative". I drive at 150 km/h relative to the asphalt of the motorway. The Porsche overtaking me drives relative to me with 80 km/h. Without this reference, the value of a speed is worthless. But in the Maxwell equations, the speed of light appears without any reference. This could be interpreted as the speed of light relative to the ether. After it was off the table, Albert Einstein concluded from the missing reference that the speed of light is always the same. No matter how fast you move. This is the key to the special theory of relativity which he published in 1905.

If one flies behind a beam of light at 99% of the speed of light, it will move away at 100% of the speed of light and not at 1%. This constancy of the speed of light has big consequences: The faster I move relative to an observer, the slower my time flows for him and the shorter my body appears to him in the direction of movement. I don't notice it myself. The changes concerning my time and extension are noticeable only by the observer. Although I notice the same changes in him. This all follows from the constancy of the speed of light. Besides, it also represents the highest speed at all. Nothing can move faster than the speed of light.

In 1905 Albert Einstein was still a clerk at the Bern Patent Office. His title was "Technical Expert 3rd Class". In addition to the special theory of relativity, he has published three other groundbreaking papers this year. And he also submitted his doctoral thesis. That is why 1905 is called Einstein's "miracle year". One of the three other papers was the explanation of the photoelectric effect. It provided an important insight for the development of quantum theory, as I will explain in the next chapter. Later, however, Einstein was its vehement adversary, considering it incomplete until the end of his life.

We are approaching the end of classical physics and the beginning of quantum theory. Although that overlaps in time. The last chapter of classical physics is the general theory of relativity. Albert Einstein published it in 1915. Quantum theory, on the other hand, as you will see in the next chapter, started as early as 1900.

The special theory of relativity fits wonderfully to the Maxwell equations, because according to them, changes in the electromagnetic fields and thus changes in the effects on other bodies always spread with the speed of light. In Newton's law of gravity, however, there is an instantaneous long-distance effect. The earth, for example, attracts the moon. If the earth would suddenly explode, the moon would notice it without delay. Newton didn't like that and in fact, it massively contradicts the special theory of relativity. In 1907 Einstein therefore began to search for an alternative, when he was still a largely unknown employee at the Bern Patent Office. Whereas in 1915, the year in which the general theory of relativity, which can also be called the theory of gravitation, was completed, he was a highly respected professor in Berlin.

The general theory of relativity consists of much more complicated mathematics than the special theory of relativity. Since Einstein was, as already stated, a better physicist than mathematician, the development of the mathematical formalism caused him great difficulties. He needed the support of mathematicians he was friends with. Einstein has made fundamental thoughts about gravity and has finally concluded that it is not a force at all. Rather, all bodies change the space and time around them. This is the famous curved space-time. It is this curvature that leads to the gravitational attraction between bodies. It represents a field, like the electromagnetic field. And all disturbances propagate at the speed of light. They're called gravitational waves. Their existence could be shown in 2015.

Albert Einstein was awarded the Nobel Prize in 1921. But not for his two theories of relativity, but for the explanation of the photoelectric effect. The two theories of relativity were never appreciated accordingly by the Nobel Prize Committee. With the prize money, he paid for his divorce, by the way.

According to classical physics (except Newton's law of gravity), all forces are transferred from one body to another through an agent. Through electromagnetic waves, for example. And the maximum transmission speed is the speed of light. This is called the "locality" principle. Because every force exerts its effect only when the transmitting agent has arrived "on-site". You already suspect it, with quantum theory this principle of locality no longer applies. It's nonlocal. This can be seen in entanglement, the most spectacular phenomenon of all. This is understandable, since he justified locality with his theories of relativity. He called the entanglement the "spooky action at a distance". For him, it was absurd and showed him that quantum theory must be incomplete. Einstein did not live anymore to see it, but entanglement could be proven experimentally meanwhile. For once he was wrong, the spooky action at a distance really exists. However, there is no hint of an explanation for it so far. Which, as with consciousness, is due to the current scientific view of the world. It is incomplete and needs to be supplemented.

Chapter 8 : Popular Experiments

Unified Field Theory

In the past, seemingly different interaction fields or forces appeared to have been unified together. For example, James Clerk Maxwell effectively combined magnetism and electricity into electromagnetism in the 1800s. In the 1940s, Quantum electrodynamics translated his electromagnetism into the terms and mathematical equations of Quantum mechanics. During the following decades, physicists successfully unified nuclear interactions, both strong and weak, along with Quantum electrodynamics to create the Standard Model of Quantum Physics.

So what is the difficulty for scientists? The problematic issue with a fully unified field theory is gravity is best explained by Einstein's theory of general relativity. But the other three fundamental interactions and their quantum mechanical nature are best explained with the Standard Model.

As we will discuss later, this illusive theory has been seen as a holy grail of sorts, thought to solve several different issues regarding the number of string theories, as well as answering the questions that result from the inability to find an equation that combines all four fundamental forces.

Black Body Radiation

In 2009 I started to look closer to our consciousness. How is it generated? That's a complete mystery. And it's unique. There are many mysteries. Such as, for example, the hitherto impossible combination of quantum theory with general relativity. Or the mystery of the origin of life and cancer. But for all these mysteries there are at least proposals for a solution. With consciousness it's different.

What is consciousness? Some say that it is what comes up when we wake up in the morning and disappears when we fall asleep in the evening. Whereby we are conscious also during sleep when we dream. The special feature of dreaming is that there is neither an input from the sensory organs nor an output to the muscles. Rather, the brain is fully occupied with itself.

Generally accepted is the definition of consciousness as an experience. One experiences what comes from the sensory organs and the body. But one also experiences memories and abstract thoughts. This is accompanied by feelings and emotions, which are also part of our consciousness. All this is meant by "experience". Whereby one usually adds that it is a "subjective" experience. Because it's not discernible to outsiders. Only I myself know of my experience. I can report on it, but it is not possible for my counterpart to judge whether what I am saying is true. Hence the addition "subjective".

Our entire existence consists completely of our consciousness. Because we know only the experience connected with it, we do not know anything else. If consciousness disappears, then the perception of our existence also disappears. Without consciousness, I don't know that I exist. Which shows that it's for us the most important phenomenon of all.

Therefore, people have always wondered how consciousness is created. The neuroscientists and philosophers have asked this question particularly intensively, and continue to do so. The neuroscientists are convinced that consciousness is created entirely by the brain. Unfortunately they cannot even begin to show how this is done. What is certain, however, is that there is a connection between consciousness and brain. Consciousness is completely dependent on brain activity. If it is strongly changed, as with deep sleep or anaesthesia, then consciousness disappears.

The neuroscientists have not yet come any closer to consciousness itself, i.e. how it comes to the experience associated with it. Nevertheless, they have made progress. In recent years they have succeeded in finding so-called correlates of consciousness. This means that certain characteristics of brain activity show that it is associated with consciousness. I'll come back to that later.

Now to the philosophers. They try to track consciousness through pure thought. Not entirely unsuccessful, as we shall see shortly.

There are two schools of thought. One is dualism, which today only plays an outsider role. It goes back to the Frenchman René Descartes (1596-1650), who was very versatile. He was a lawyer, mathematician, physicist and philosopher. But also a soldier in the Thirty Years' War. Descartes, in any case, said that mind and matter are two completely different things. He didn't distinguish between mind and consciousness. That's still common today. According to Descartes, matter is expanded, so it can be perceived with the senses. Mind, on the other hand, is not extended, so one cannot perceive it directly. That's dualism. For its followers, consciousness is completely detached from all matter, especially the brain. The big problem with dualism is that it cannot explain how consciousness interacts with the brain. So why there is the observed close relationship between consciousness and brain. This is the reason why dualism today only plays an outsider role.

The second school of thought is monism. In its standard version, it says that consciousness is a property of matter. And of exactly the matter as it is described by physics. The standard version of monism is therefore also called physicalism. Behind this is the idea that the entire reality is captured by physics. Including consciousness. However, physicalism also has a huge problem. For it cannot begin to explain how the subjective experience of consciousness emerges from physics, for example the experience of the red of a sunset. This is what the Australian philosopher David Chalmers (born 1966) calls the "hard problem" [18].

But there is a version of monism that provides a small glimmer of hope. Let's say it provides the idea for a solution. This is the so-called dual-aspect monism.

Let's look at an electron. What "is" an electron? Physics provides its unchangeable properties, which are electric charge, mass and spin. When physics speaks of an electron, it means the bundle of its unchangeable properties. But they only show what an electron "does", i.e. how it affects other objects and vice versa is influenced by them. This is quite reasonable, because we can only see what matter "does". Therefore, this is exactly what can be extracted from matter with the methods of physics. But many philosophers and also some physicists suspect that this cannot be the whole truth. Maybe physics delivers with respect to matter only the surface of a substance. Let us call the surface the external nature of matter and the substance its inner nature. It represents what matter "is". Concerning the substance, physics must remain silent. But it is more fundamental because without the substance there is no surface. Therefore and that's very important is the external nature of matter created by its inner nature.

The British philosopher, mathematician and logician Bertrand Russell (1872 - 1970) also assumed that matter has an inner nature. And he went one step further, because he brought consciousness into play. In

1950, in his essay "Mind and Matter", he said the following: "We know nothing about the inner nature of things. Unless our consciousness tells us something about it."

This is precisely the basis of the dual-aspect monism. There is indeed only matter, hence the expression "monism". But there are two aspects to it. One is physical in the form of the external nature. The other is the inner nature, to which physics cannot say anything, but from which the physical aspect of matter emerges. The inner nature is closely connected with consciousness, this is the basic statement of the dual-aspect monism. Further above I have said that the standard version of monism is physicalism. It assumes that the entire reality is captured by physics. This, of course, also includes consciousness, but so far physics has not been able to say anything to it, this is the "hard problem". The dual-aspect monism is different. Not everything is physics, it provides only one aspect of matter, namely its external nature. There is another which is non-physical, that's the inner nature of matter. It provides the basis for consciousness.

There are two variants of the dual-aspect monism, one radical and one moderate. For the radical variant, the inner nature of matter already represents consciousness. The experience connected with it thus consists of the inner nature. The decisive question, however, is whether the radical variant is in harmony with human consciousness. The inner nature of each atom and molecule of our body then contributes to our experience, provides a micro-experience. There are two problems. Firstly, we have a structured but uniform experience. So how are the countless micro-experiences united? This is called the combination problem and its solution is completely unclear. Secondly, our experience is focused on the brain. What suppresses the rest of the body? Here, too, there is no hint of an explanation. Let's call it the dominance problem.

For the moderate variant, the inner nature of matter consists of proto-consciousness, so it is a preliminary stage of consciousness. But it remains unclear how proto-consciousness becomes consciousness. Nevertheless, this approach is much better in principle. Because the experience is not simply "there" as with the radical variant. Rather, it is generated by a process that makes consciousness out of proto-consciousness and only this leads to an experience. Although the process is still unknown, it is conceivable that it can be linked to the processes in the body, including the brain. And the link can maybe solve the combination and the dominance problem. Both problems are also present in the moderate variant.

The solution of the mystery of entanglement

As I said earlier, in 2009 I started to deal with our consciousness. And, of course, I wanted to show how it is created. The moderate dual-aspect monism became my starting point. But the existence of the inner nature has so far been speculative. So my first step had to be to show that the inner nature really exists. There is an approach: If it really exists, then one must assume that the external nature is created by it. This establishes a connection between inner nature and physics. Could this not be used for an indirect validation? That means that a fundamental mystery has to be solved with the inner nature. For this, I chose entanglement, next to consciousness the most mysterious enigma ever.

It was immediately clear to me that for solving the mystery of entanglement, the spatial world and the presumed non-spatial world must be intimately linked. But what does the connection look like? An idea came to my mind: The inner nature is not in the spatial but in the non-spatial world. And since it creates the external nature that is of course in the spatial world, both worlds must be closely connected. For wherever there is an elementary particle, atom or molecule in the spatial world, the non-spatial world is connected to the spatial world in the form of the corresponding inner nature.

I also gave a name to the non-spatial world, I consequently call it the inner world.

To avoid any confusion, first a few words about the designations: When I speak of an elementary particle, atom or molecule in the following, I always mean its external nature, i.e. what physics understands by it. When I speak of the inner nature, I say that explicitly.

Each type of elementary particle, atom or molecule has its own inner nature, which only exists once. This is easy to understand because in the non-spatial inner world there is no spatial separation. Consequently, there cannot be two or more specimens of an inner nature. However, there are of course any number of external natures, i.e. any number of elementary particles, atoms and molecules of any kind. This works because, for example, the inner nature of the electron can connect to the spatial world at any number of locations. And at any location where that happens, an electron appears.

With this concept, as I shall now explain, the mystery of entanglement can be solved.

Let's take a look at two quantum objects A and B, both in a superposition state of black and white (of course there are no real colors in the quantum world, they stand for abstract properties). A and B merge to form a new composite quantum object, let's call it AB. The special feature of AB is that it can have only the color gray. This means that there are only two possible states for AB: A = black and B = white or A = white and B = black. This means that the color of A determines the color of B and vice versa. That's when A and B are connected. But it is still valid even if A and B are separated and light-years away from each other. This is exactly the phenomenon of entanglement.

Let us now assume that A is at location x and B is at location y, many light-years away. We are talking about the physical versions, i.e. the external natures. But there are also the inner natures. The entanglement causes the inner natures of A and B not to be separated, but to form a single inner nature. The mystery of entanglement is such a big mystery because for physics, only A exists at location x and only B at location y, which is many light years away, without any obvious connection between them. However, this is not the whole truth, for there is also the common inner nature. Both at location x and at location y it connects to the spatial world. At x it produces A and at y it produces B. Due to the common inner nature, A and B are thus connected to each other, of course independent of their distance.

Now a measurement is performed on A that changes A. It is no longer in a superposition state of black and white, but either black or white. Let's assume it's black. But the measurement has also another effect, namely the formation of a new inner nature, consisting of the common inner nature of A and B and the inner nature of the measuring device. The new inner nature also contains the information about the measurement result. But it includes A and B. Therefore the information is not A = black, but A = black and B = white. The new inner nature is connected to the spatial world both at location x and at location y. At y it causes a changed B to be produced, which corresponds to the new inner nature. It therefore loses its superposition state of black and white and becomes white. And that, of course, immediately. What explains the phenomenon of entanglement.

This justifies both the assumption of the existence of the inner world and the inner nature of matter. One can therefore say that my concept of the inner world is the preliminary stage of a theory. Which means that it is a hypothesis. Let's call it the hypothesis of the inner world (HIW).

I was then able to move on to the second step. My starting point for the explanation of consciousness was the moderate variant of the dual-aspect monism. So I had to show what the process looks like that makes consciousness out of the inner nature that is "only" proto-consciousness in the moderate variant. For this, we first take a closer look at our consciousness.

Photoelectric Effect

It was known that even a very large amount of long-wave red light (about 650-700nm) could not liberate electrons from atomic binding, while even the relatively small power of violet light (about 400nm) and ultraviolet radiation (400 to 10nm) could liberate electrons effectively, even if the total energy of violet light or UV radiation is hundreds of times lower than red light energy. So, the total radiation power was not the issue. Violet light and UV radiation had to have some other advantage, although it was logical to assume that in order to liberate electrons from atomic binding, it was necessary to apply as much energy as possible to these electrons. If the pinhole is small, while it will determine the position of a particle with higher accuracy, it will also produce a relatively broad diffraction **pattern** that renders the position of the photon on the screen uncertain. We can know in a relatively precise manner where the photon or matter particle went through the piece of paper, inside the space constrained by the pinhole's aperture. But it will be displaced laterally on the screen anyway due to diffraction and interference phenomena.

Einstein pointed out that violet light and UV radiation has a great advantage over the red light in Planck's model because they had a greater power of individual "units" of radiation – Planck's quanta – than red light. Einstein also had to assume that in order to break away from the atom, the electron, rather than gradually accumulating the energy of individual small quanta of light or absorbing many small quanta simultaneously, was capable of absorbing, for some reason, no more than a single quantum with enough energy to allow the electron to instantly overcome stretching of the nucleus.

So, each quantum of red light has not enough energy to cause the effect, whereas each much-larger UV radiation quantum has enough energy to liberate one electron at a time. And the larger the quantum of UV radiation is, the higher the electron's speed would be after leaving the atom, as there will be more energy left over after the liberation of the electron. This theory perfectly matches observable reality.

Chapter 9 : Superconductors

No resistance

When resistance disappears? The answer to this question was Kamerlingh Onnes early as 1914. He proposed a very ingenious method of measuring the resistance. The experimental scheme looks quite simple. Lead wires from the coil in a cryostat omitted - an apparatus for carrying out experiments at low temperatures. Cooled helium coil is in the superconducting state. In this case, the current flowing through the coil, creating a magnetic field around it, which can be easily detected by the deviation of the magnetic needle located outside the cryostat. Then close the key, so that now the superconducting stroke was short-circuited. A compass needle, however, was diverted, indicating the presence of current in the coil, is already disconnected from a current source. Watching the arrow for a few hours (until evaporate the helium from the vessel), Onnes had not noticed the slightest change in the deflection direction.

According to results of the experiment Onnes concluded that the resistance of the superconducting lead wire of at least 10 11 times lower than its resistance in the normal state. Subsequently conducting similar experiments, it was found that the current decay time exceeds many years, and this indicated that the superconductor resistivity less than 10 25 ohm · m. Comparing this with resistivity of copper at room temperature, 1.55 x 10 -8 ohm-m - the difference is so large that one can safely assume that: the resistance of the superconductor is zero, it is difficult to actually call another monitor and modify the physical quantity which would be addressed in the same "round zero" as the conductor resistance below the critical temperature.

Recall known from school physics course Joule - Lenz: the current I flowing through the conductor with a resistance R it generates heat. At this consumed power P = I 2 R. As little resistance to metals, but it often limits the technical possibilities of different devices. Heated wires, cables, machines, apparatus, therefore, millions of kilowatts of electricity literally thrown to the wind. Heating limits the throughput power, the power of electric cars. Thus, in particular, is the case with electromagnets. Obtaining strong magnetic fields requires a large current, which leads to the release of an enormous amount of heat in the windings of the solenoid. But superconducting circuit remains cold, the current will circulate without damping - impedance equal to zero, no power losses.

Since the electric resistance is zero, the excited current of a superconducting ring will exist indefinitely. The electric current in this case resembles the current produced by the electron orbit in the Bohr atom: it is like a very large Bohr orbit. Persistent current and the magnetic field generated by it can not have an arbitrary value, they are quantized so that the magnetic flux penetrating the ring takes values that are multiples of the elementary flux quantum F on = h / (2e) = 2.07 10 15 Wb (h - Planck's constant).

Unlike electrons in atoms and other microparticles, whose behavior is described by quantum theory, superconductivity - macroscopic quantum phenomenon. Indeed, the length of the superconducting wire through which flows persistent current can reach many meters or even kilometers. Thus carriers it describes a single wave function. This is not the only macroscopic quantum phenomena. Another example is a superfluid liquid helium or substance neutron stars.

In 1913 Kamerlingh Onnes is proposing to build a powerful electromagnet coils of superconducting material. Such a magnet would not consume electricity, and with it could receive superstrong magnetic field. If so ...

Once tried to pass through the superconductor significant current, the superconductivity disappears. Soon it turned out that a weak magnetic field also destroys superconductivity. The existence of critical values of temperature, current and magnetic induction severely limiting the practical possibilities of superconductors.

Electrical resistance superconductors

No experimental methods fundamentally impossible to prove that any value, in particular the electrical resistance is zero. It can only show that it is less than a certain value determined by the measurement accuracy.

The most accurate method of measuring low impedances consists in measuring the current decay time induced in the closed circuit of the test material. Decrease in current time energy LI 2 /2 (L - inductance loop rate) consumed for Joule heat: here integrating,(I_0 - current value at t = 0, R - resistance of the circuit).

The current decays exponentially with time, and the attenuation rate (at a given L) is determined by the electrical resistance.

For small R formula (8) can be written as here dI- current change during Δt. Experiments conducted using a thin-walled superconducting cylinders with extremely small values of L, showed that the superconducting current is constant (with accuracy) within a few years. It followed that the resistivity in the superconducting state is less than $4 \cdot 10^{25}$ ohm m, or more than 10^{17} times less than the resistance of copper at room temperature. Since the possible decay time comparable to the time the existence of mankind, we can assume that R DC in the superconducting state is zero.

Thus, the superconducting current - it is only in the nature of real-life example of perpetual motion on the macroscopic scale!

When R = 0 the potential difference V = IR on any segment of the superconductor and hence the electric field E inside the superconductor is zero. The electrons which create a current in the superconductor, moving at a constant speed without being scattered by the thermal vibrations of the crystal lattice atoms and its irregularities. Note that if E is not equal to zero, electrons carrying the superconducting current to accelerated without limit and the current could reach an infinitely large value, which is physically impossible. To create a superconducting current, it is necessary only to accelerate the electrons up to a certain speed directional movement (expending this energy), and further the current is constant, without borrowing power from an external source (as opposed to a conventional current conductors).

The situation changes if the superconductor is applied to the variable potential difference creates a variable superconducting current. During each period the current changes direction. Consequently, in the superconductor must exist an electric field which periodically slows down the superconducting electrons and accelerates them in the opposite direction? Since it consumes energy from an external source, the electrical resistance of alternating current in a superconducting state is zero. However, because the electron mass is very small, power loss at frequencies less than 10 10 - 10 11 Hz are negligible.

Note that the presence of the Meissner effect (vanishing magnetic induction inside the superconductor material) superconducting current flows only in a thin layer on the surface whose thickness is determined by the depth of penetration L of the magnetic field into the superconductor, and at high frequencies where the depth of the surface layer, in which penetrates an alternating electromagnetic field becomes smaller l - even thinner layer.

The intermediate state in the destruction of superconductivity current

Upon reaching the critical magnetic field of superconductivity abruptly collapses and wholly sample passes to the normal state. This is true even when the external magnetic field has the same value at any point on the sample surface. This simple situation can be realized, particularly for very long and thin cylinder with its axis parallel to the field.

If the sample has a different shape, the picture transition in a normal state looks a lot more complicated. With increasing field comes a point when it becomes critical in any one location of the sample surface. If the sample is a sphere, the magnetic field ejection leads to thickening of the lines of force in the vicinity of its equator. Such common field is the result of blending to a uniform external magnetic field induction B 0 magnetic field generated by currents filmed.

Clearly, the distribution of magnetic field lines due to the geometry of the sample. For simple bodies characterize this effect can be a single number, the so-called expansion coefficient. If, for example, the body has the shape of an ellipsoid, one of which is directed along the axes of the field, the field at its equator becomes equal to the critical when condition B 0 = B from (1-N). × When a certain coefficient N demagnetization field can be determined on the equator. If the pinhole is small, while it will determine the position of a particle with higher accuracy, it will also produce a relatively broad diffraction **pattern** that renders the position of the photon on the screen uncertain. We can know in a relatively precise manner where the photon or matter particle went through the piece of paper, inside the space constrained by the pinhole's aperture. But it will be displaced laterally on the screen anyway due to diffraction and interference phenomena.

For a sphere, for example, $N = \frac{1}{3}$ at the equator so that its magnetic field become equal to the critical when induction $B\ 0 = \frac{2}{3}V$ with . With further increase of the field at the equator should superconductivity break down. However, all the ball can not go back to normal, as in this case, the field would penetrate into the interior of the sphere and would be equal to the external, that is, the field proved to ba less critical. There comes a partial destruction of superconductivity - pattern stratified into normal and superconducting regions. A state where the specimen contains normal and superconducting regions, called intermediate.

The theory of the intermediate state was developed by Landau. According to this theory in the range of magnetic fields with induction $B\ 1 < B\ 0 < B$ with ($B\ 1$ - Induction of the external magnetic field, the moment when in any location on the surface of the field induction B reaches the value a). Superconducting and normal regions exist, forming a plurality of alternating zones of different electrical conductivity. The real picture is much more complicated. The structure of the intermediate state is obtained in the study of the tin ball (superconducting regions are shaded). And N - - regions continuously changes the ratio of the S. With increasing field superconducting phase "melts" due to growth N - regions and the induction B = B to disappear completely. And all this due to the formation and disappearance of the boundaries between the S - and N - regions. A formation of any interface between two different states to be associated with the extra energy. This surface energy plays a very important role and is an important factor. From it, in particular, depends on the type of superconductor.

The normal area is equal to the critical magnetic field (or more). On the border there is a sharp transition from completely normal to a fully superconducting. The magnetic field l penetrates a distance into the superconducting region, and the number of superconducting electrons n s per unit volume is increased slowly over a distance equal to a characteristic length, which. x called coherence length

The penetration depth, l is of the order of 10 -5 ... 10 - 6 cm, the coherence length for pure metals, the estimated A.Pipparda English physicist must be of the order of 10 -4 cm. As shown by Soviet Physics

V.L.Gizburg and LD .Landau, the surface energy is positive, ecli x \ l ratio of 2 Ö less than 1 \ " 0.7. This case is realized in materials, which are called type-I superconductors.

In superconductors surface energy is positive, that is, in the normal state is higher than in the superconducting state. If the thickness of such material normal zone occurs, then the boundary between normal and superconducting phases required some expenditure of energy. This explains the reason for separation of a superconductor in the intermediate state is only a finite number of zones. The dimensions of S - and N - regions may be on the order of a millimeter, and they may be visible to the naked eye, covering the surface of the sample and the superconducting thin magnetic (diamagnetic) powder. Magnetic powders are attracted to the field and positioned at the output of the normal layers.

Now for the type-II superconductors. Intermediate state corresponds to the situation when the separation l <. x In heterogeneous metals in the presence of impurities is not the case. The collision of electrons with atoms of impurities may reduce the coherence length. x In materials such as alloys, the coherence length is less than, and sometimes substantially - hundreds of times than the depth of penetration. Thus II superconductors - this alloys and metals with impurities. In type II superconductors I surface energy is negative (<), x is why the creation of the interface between the phases associated with the release of some energy. They energetically favorable to flow into a volume of the external magnetic current. The substance thus breaks down into a mix of small superconducting and normal regions, the boundaries of which are parallel to the applied field. Such a condition is called mixed.

superconductors in a magnetic field

The fact that a magnetic field greater than a threshold or critical value, the superconductivity disappears completely certain. Even if some metal would lose resistance when cooled, it can not go back to normal, once an external magnetic field. In this case too, the metal is recovered about the resistance that it was at a temperature above the temperature T csuperconducting transition. Needless critical field with the magnetic induction B with temperature dependent: induction is equal to zero at $T = T_c$ and the temperature rises tending to zero. For many metals in dependence induction with temperature is similar.

Consider now the behavior of a perfect conductor (i.e., conductor deprived of resistance in different environments). Such a conductor when cooled below a critical temperature, the electrical conductivity becomes infinite. This property has allowed to consider a superconductor perfect conductor.

The magnetic properties of an ideal conductor emerged from the law of induction - Faraday and conditions of infinite conductivity. Assume that a metal transition to the superconducting state occurs in the absence of a magnetic field and an external magnetic field is applied only after the disappearance of resistance. It does not need any subtle experiments to ensure that the magnetic field inside the conductor does not penetrate. Indeed, when metal enters the magnetic field, on its surface due to electromagnetic induction occur is not damped closed currents (the number of call screening) creating its magnetic field induction whose modulo is, external magnetic field and the direction vectors of magnetic induction of the field opposed. As a result of the induction of the resultant magnetic field is zero.

A situation arises in which the metal as it prevents the penetration of the magnetic field, that is behaves as a diamagnetic. Now, if the external magnetic field is removed, the pattern will be in a not magnetized.

Now put in the magnetic field of the metal in the normal state, and then cool it to he transferred to the superconducting state. The disappearance of the electric resistance should not influence the magnetization of the sample, and therefore the magnetic flux distribution will not change it. If now the applied magnetic field is removed, the change in the flux of the external magnetic field through the sample volume will result (according to the law of induction) in the appearance of persistent flows, a magnetic field which exactly compensate for the change in the external magnetic field. As a result, the captured field will not be able to escape: it will be "frozen" in the amount of sample and remain there in a kind of trap.

As can be seen the magnetic properties of an ideal conductor depends on how it gets to the magnetic field. In fact, at the end of these two operations - the application field and reducing - the metal is in the same conditions - at the same temperature and zero external magnetic field. However, the magnetic induction-metal sample in both cases is quite different - zero in the first case and the end, depending on the source field in the second.

Josephson effect

Physical objects in which the Josephson effect takes place, now called Josephson junctions or Josephson junctions or Josephson elements. In order to imagine the role played by Josephson elements in superconducting electronics, it is possible to draw a parallel between them and the semiconductor p-n-junctions (diodes, transistors) - element base conventional semiconductor electronics.

Josephson junctions are some weak electrical communication between two superconductors. In fact, this connection can be done in several ways. The most commonly used types in practice weak link - is: 1) tunnel junctions, in which the bond between the two film superconductor is carried out through a very thin (tens of angstroms) insulating layer between them - SIS-structure; 2) "sandwiches" - two film superconductor interacting through a thin (hundreds Angstroms) layer of a normal metal therebetween - SNS-structure; 3) the structure of the bridge type, which are narrow superconducting bridge (bridge) of limited length between two massive superconducting electrodes.

Bearers overcurrent superconductors at T = 0 K are all conduction electrons n (0) (electron density). When the temperature rises appear elementary excitation (normal electrons), so that the concentration of n s of the superconducting electrons at a temperature T

n s (T) = f (0) -n n (T) ,

where n n (T) - the concentration of electrons at a normal temperature T. In the Bardeen-Cooper-Schrieffer (BCS) for T → T c (critical temperature)

p s (T) ≈ Δ 2 (T)

where 2 D (T) - the width of the energy gap in the spectrum of the superconductor. All superconducting electrons form pairs associated state, known as Cooper pairs of electrons.

Cooper pair combines two electrons with opposite spins and pulses, and therefore has a zero net spin. Unlike normal electrons having a spin of 1/2, and therefore obey Fermi-Dirac statistics, Cooper pairs obey Bose-Einstein statistics and condensed at one, the lower energy level. A characteristic feature of Cooper pairs is their relatively large size (about 1 micron) is much greater than the average distance between pairs (of the order of the interatomic distances). Such a strong spatial overlap pairs means that all of the (condensation) of the Cooper pairs is coherent, that is described in quantum mechanics, the wave function of a single W = .DELTA.E ix . Here A - amplitude of the wave function, the square of which characterizes the concentration of Cooper pairs, h - the phase of the wave function, i - imaginary unit, P - -1. In the case of normal electrons, which are fermions, the Pauli exclusion principle, the electron energy is never exactly equal to each other. Therefore, from the Schrodinger equation for these particles it follows that the phase velocity dq / dt of the wave functions of electrons normal differ thus phase h are uniformly distributed in the trigonometric circle, and the summation over all particles explicit dependence on h disappears.

The presence of a weak electric connection between the superconducting electrodes due to poor overlap of the wave functions of the Cooper pairs of electrodes, whereby such contact is also superconducting, but the density of the critical current value is much (by several orders of magnitude) smaller than the critical current density of the electrodes $j_c \approx 10^8 \, A/cm^2$. For tunneling structures and structures of sandwich-type critical current density of Josephson junctions-ing is typically in diapazone j_{jc} from 10^1 to $10^4 \, A/cm^2$, and their area SB within modern technology can be made from a few hundred to a few square microns. Therefore, the critical current of the Josephson element $I_c = j_{jc} \cdot S$ may be from a few milliamperes to a few microamperes.

In general it can be noted three consequences manifestation of quantum coherence of Bose condensation of Cooper pairs in the macroscopic scale:

1) the mere presence of an overcurrent in the superconductor,

2) the effect of weak links in Josephson superconductors and finally

3) the effect of the quantization of the magnetic flux.

The quantity of permanent overcurrent through the Josephson junction is a periodic function of the phase difference of the wave functions of the electrodes $i = h_1 - h_2$, called Josephson phase.

In the absence of current through the Josephson element n = 0 (up to 2rn), and when the maximum supercurrent flow equal to I c , the Josephson phase q = p / 2. When the flow of direct current I> I with the voltage across the junction equal to zero. This phenomenon is called the Josephson effect.

Ac Josephson effect (Dch / dt ≠ 0) occurs when, for example, through the Josephson element is passed a current I> I c . In this case, the transfer of current I through the Josephson junction except supercurrent I s will also participate normal component I n , which is the normal current of electrons n n (T). Thus, I = I s + I n . The flow of the normal and hence a dissipative component of the current causes the appearance at the transition of the Josephson-uted voltage drop

V = Q n R n ,

wherein R n - so-called normal resistance. In view of the fundamental relation Josephson in this case will take place unlimited increase (or decrease, if V < 0) Josephson phase n and, hence, the periodic variation in time of the overcurrent I s . Thus, when passing through the Josephson element DC | D>> I c , this current is carried by two components of the current I s and I n , which, according to (3) oscillate (in the pro-tivofaze) in time with a frequency proportional to the DC component V voltage drop across the Josephson junction:

The voltage across the Josephson element of V (t) = I n (t) R n will also oscillate in time at a frequency w and this process is called Josephson generation. This state of the Josephson junction is called resistive. It should be emphasized that despite the presence of the voltage drop across a Josephson junction, superconductivity electrodes forming a Josephson element is stored in the resistive state.

If Josephson element has appreciable self-capacitance C (e.g., tunneling Josephson junction), the resistive state of current therethrough will be the sum of three components: I s , I n and a capacitive component of the current (bias current)

The simplest model of Josephson elements well type structure describing the bridge and S-N-S, a resistive model, wherein normal resistance R n is constant not depending on the voltage V.

Tunnel effects

In 1962, an article appeared before anyone unknown author B. Josephson, which theoretically predicted the existence of two extraordinary effects: steady and unsteady. Josephson theoretically studied the tunneling of Cooper pairs from one superconductor to another through any barrier. Before proceeding to the first Josephson effect, briefly the tunneling of electrons between the two metal parts separated by a thin dielectric layer.

The tunnel effect known in physics for a long time. The tunnel effect - this is a typical problem of quantum mechanics. Particle (for example, an electron in the metal) approaches the barrier (eg, a dielectric layer), to overcome which she classical ideas can not, as its kinetic energy is insufficient, although the area of the barrier it with his kinetic energy could well exist . On the contrary, according to quantum mechanics, the barrier passage possible. The particle may have a chance, as it were, to pass through the tunnel through classically forbidden region where its potential energy would be more like a full, ie the classical kinetic energy as it is negative. In fact, from the point of view of quantum mechanics for the microparticles (electron) holds uncertainty relation $\Delta h \Delta r > h$ (x - coordinate of the particle, p - its pulse). When a small uncertainty of its coordinates in a dielectric $\Delta h = d$ (d - thickness of the dielectric layer) leads to a large uncertainty its pulse $Dp \geq h / \Delta x$, and consequently, the kinetic energy $p 2 / (2m)$ (m - mass of particles), the energy conservation law is not violated. Experience shows that indeed between two metal electrodes separated by a thin insulating layer (tunnel barrier), electric current can flow the greater, the thinner the dielectric layer.

Chapter 10 : Going 3D with Quantum physics

I will derive equations in a method foreign from what you were taught in graduate school. Some of my logical explanations may appear a bit naïve and almost an assumptive, but this approach helps newbies understand my work. I will purposely simplify every equation and remove as many foreign symbols as possible to eliminate confusion and clutter. Ultimately this is a brief presentation supporting my interpretation and derivation of an alternative theoretical TOE model. It is not intended to explain the Standard Model's interpretation. So hold on to your seats and enjoy the ride!

There is an old saying, seeing is believing, but that's not entirely true in Quantum Mechanics. If the pinhole is small, while it will determine the position of a particle with higher accuracy, it will also produce a relatively broad diffraction **pattern** that renders the position of the photon on the screen uncertain. We can know in a relatively precise manner where the photon or matter particle went through the piece of paper, inside the space constrained by the pinhole's aperture. But it will be displaced laterally on the screen anyway due to diffraction and interference phenomena.

To this, memorizing is not learning, calculating is not understanding, and describing is not explaining. If the Standard Model does nothing more than name observations in an experiment, it is totally useless in its ability to understand nature. Birds do not fly because they have wings. Just ask Icarus. Particle physicists will eventually become the chemists of atoms, which provide a necessary contribution, but we will not find an explanation of anything by grouping and cataloging particle fragments. Naming something and describing its purpose cannot define how that something functions. If we want to truly understand how, we need to study why... not what!

Sometimes when you listen to a conference or if you decide to read a paper on physics, you will notice the many hand-waving tricks designed to misdirect your attention and gain your influence. When magicians reveal their illusions, they always arrive at a point in the trick where they do something unnecessary to divert your attention long enough to keep you amazed while they perform the essential aspects they don't want you to observe. Towards the end of the show they raise their hands, modify voice inflections, and arch their eyebrows in total amazement to communicate to you they have accomplished what they said they've accomplished. The audience is confused because internally they are reasonably skeptical and do not believe the illusion, but they're convinced by the magician's buffoonery to believe it. This is the magician's only trick...to convince you to believe something that is not true. These illusions are great at exhibiting entertainment but far from producing enlightenment.

Understanding the many aspects of the illusion will help you avoid confusion;

1. **Hippopotomonstrosesquipedaliophobia:** Beware when a speaker or author rambles using only big words you don't understand. Chances are the only thing he knows more than you is a bunch of fancy words.

2. **Fake Facts:** When a speaker or author excessively use equations to communicate an idea. Chances are the equations are wrong or presented to support their fabricated

concepts. Always look for the new mathematical term that conveniently pops out from nowhere like a trap door.

3. **Intractability:** When a speaker or author pretends they can't explain it any simpler. How hard is it to explain what happens at a surprised birthday party? If you truly understand something you can break it down to a 5th grader. Simply ask them to. If they can't, it's a sure sign they do not know what they are talking about.

4. **Intimidation:** They dazzle you with large expensive equipment and mention big numbers (costs, speeds, time scales, etc...). This is to intimidate you as an individual. Make these larger than life things irrelevant or replace the large number with the words "a number".

5. **Appearance:** They use an accent and wear eccentric costumes. A business suit or a lab coat is a costume, especially if it is done at an informal event. Don't judge a book by its cover. Our eyes are trained to prejudice our mind and the speaker knows it.

6. **Accreditation:** This is usually done in the beginning during the introduction. Don't get hyped. Someone's past accolades has nothing to do with what they are about to present. It is best ignoring the introduction. At the end of a long tenure, you would probably go on a few speaking engagements if the price is right. Depending on your needs you may say just about anything also...if the price is right.

I understand this may seem harsh to some, but nature offers a fair playing field. You must stand on the merit of your present presentation. To do this, teach as if it's for 5th graders. Do not rely on special effects, intimidation, and human psychology for persuasion. Present only the facts...just the facts!

If we do not know something, it's ok to say, "I don't know?" This is what keeps us in business...our ignorance! I mean that in the most accurate sense. Our ignorance is not forbidden it's formidable. We are conducting research experiments because we do not know the answer to a particular question and the research helps us figure it out. Computer programmers debug computers by looking for unexplained problems. Every error provides a clue to a better understanding. We may resort to blaming children if our ego's get in the way. In fact, I do it all the time. I say "let me assure you, all of my "expensive" research helps children all around the world resolve their instinctive curiosities about the universe in which they were born". That is a very true statement and it helps my ego. Actually, only adults accept things they don't understand. Children do not instinctively know how not to ask questions, that comes with maturity. So if your child persistently asks why, chances are, that child will become great scientist or salesman...hopefully not both.

What if we spend a lot of money on an experiment and it does not go as planned? Every executed experiment goes as planned. Either it proves or disproves something. Experimenting is like hand feeding a trunk-less elephant. You're not really sure which side to insert the food, but you learn quickly! It is always best to support another theory, just in case the first option does not work the way you've planned.

Remember our experiments test our mathematical models based on our interpretations of them. They do not offer guarantees. This does not mean our accrued summit of work is futile or nonsense because we do not have an answer. It means we did not reach the correct conclusion. Any idiot can make a guess or a wrong conclusion. We are disciplined to find the right answers using the evidence of quantifiable facts. Until we find the right answers...we must continue iterating until we do. Once achieved, we must continuously investigate what we believe. Together with taxes, it is simply a part of our current existence.

If we are lucky, we may accumulate all of the necessary numbers called in a lottery. This is a great achievement, but the numbers must be arranged in the right position to actually win the jackpot. The Standard Model is not the TOE, it's not even wrong to be close. The advancements made in the twentieth century however have shown why approaching this one end is not the favored approach. It has also led to new technologies needed to disprove or prove any future TOE models.

So far, we humans have discovered the universe's hidden code, deciphered the symbols, and ruled out many interpretations. Now we must organize the symbols to interpret the code correctly independent and related forms, like children born to parents; each offspring is a derived copy of the original individual parent. The equations we use are simply scrambled and therefore do not make uniformed logical sense to anyone as a TOE.

Chapter 11 : Angular Momentum on Quantum Level

The most often cited example of angular momentum is a skater that spins rotates in a circle and drawing his/her arms closer to the body and spins faster.

But the best examples of angular momentum are the planets and stars and galaxies and satellites. For angular momentum and gravity are tied up together.

In Old Physics, they thought angular momentum was the units:

Angular Momentum = kg*m^2/s which is kilograms*meters^2 / seconds. That is mass*area / time. Trouble with Old Physics angular momentum is they forgot a important term of electric current (i).

The true equation of angular momentum needs a electric current (i) term in it.

angular momentum = kg*m^2/(i)*s

You see that extra term (i) in the denominator? This is the important term they missed in 20th century physics.

Angular momentum L = kg*meter^2/((i)*seconds) where electric current "i" the term that every physicist of the 20th century missed.

Or, you can do Angular Momentum as represented by geometry of Kepler's three laws of motion:

1) the orbits of planets is an ellipse with the Sun at a focal point.

2) planets sweep out equal areas in equal time.

3) the square of orbit period is proportional to cube of semimajor axis.

Angular momentum in classical mechanics

Before focusing on the mysteries and paradoxes of the quantum world, we must learn a couple of elementary concepts – such as angular momentum and spin – that we will find throughout the rest of this book. So, what is angular momentum? In CP it is not very hard to understand it. It is the amount of rotational momentum, the quantity of motion of rotating bodies. Therefore, let us first look at the notion of the angular momentum of a single particle rotating around a center.

Where is the position vector of the particle relative to the rotational origin, and the cross denotes the vector or cross-product (a sort of multiplication for vectors). Please note that both the linear and the angular momentum are vector quantities, which means they have a magnitude and direction. Letters representing vectors are written with an arrow or in bold.

(Here, we use the latter.) The angular momentum vector **L** is perpendicular to the **r x p** plane. Its magnitude (length) is a measure of the rotational speed.

In the event of a body made of n particles rotating around a center external to itself (that is, orbiting around a center of origin), the` **orbital angular momentum**', can be obtained by determining the sum of the linear momenta of each of its constituent particles the sum over all the particles of masses times the respective rotational velocity for each, as:

$$= \Sigma \times$$

. ,

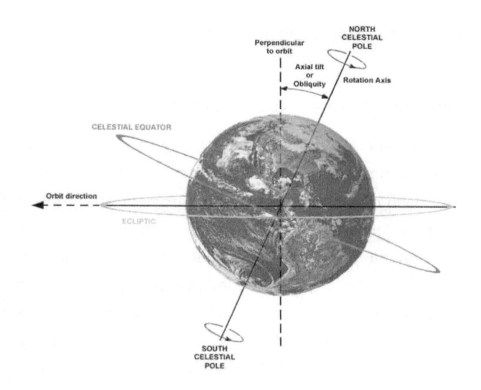

again with as the i-th particle's distance from that axis and n as the number of particles.

However, an extended body can also rotate around itself. A typical example is that of the Earth, which not only orbits around the Sun but (in case you didn't notice) spins once around its polar axis approximately every 24 hours.

In this case, for bodies whose interiors have a complex mass distribution, like the Earth (the density distribution inside the earth can be quite complicated), one must calculate the so-called **'moment of inertia'** of a body.

How our human consciousness comes into being

First a short repetition of the basic principle of consciousness: Every forming atom or molecule has a brief flash of consciousness at the moment of formation. It experiences the information provided by the corresponding inner nature that emerges in it and organizes the spatial relationships between its components. Every flash of consciousness arises exactly at the boundary between the spatial and the non-spatial inner world. It therefore has both a spatial and a non-spatial character. As we will soon see, this is crucial for the explanation of human consciousness.

Everywhere in our body molecules are constantly formed in immeasurable numbers. Because every biochemical reaction means the formation of molecules. Each of these formations leads to a flash of consciousness, combined with a micro-experience. So we should have a full-body consciousness. But our consciousness is focused on the brain. That's the dominance problem. The combination problem must also be solved: How the many micro-experiences unite to our uniform human experience. In fact, the combination problem is divided into two. Here is a simple example: You see a red ball and a blue cube lying on the table in front of you. Then four modules become active in the brain: "red", "blue", "ball" and "cube". "red" and "ball" are integrated, the "red ball" is experienced. Although the two modules are spatially separated from each other. The same applies to "blue" and "cube". But there is also differentiation because the red ball and the blue cube are perceived separately. Therefore, I just said that the combination problem is divided into two. Firstly, it is necessary to clarify how integration is achieved and secondly, to show what brings about differentiation. So we need to solve three problems in total.

In the last section, we got to know the consciousness network. Its decisive characteristic is the timed firing of its neurons. As I will now explain, this, in combination with the basic principle of consciousness, is the key to solving all three problems.

Undoubtedly, the firing of neurons plays an important role in the emergence of consciousness from the consciousness network. So we have to look for molecules whose formation is closely related to firing.

There's something that comes to mind: the synapses. They are located in the connections between the neurons and are small thickenings with a gap. There is a huge number of them, on a neuron come about 1000 synapses.

If an electrical impulse arrives at one of them, it cannot pass it due to the gap. However, it leads to the release of small molecules, the so-called neurotransmitters. They cross the gap and dock on the other side to huge proteins, the receptor molecules. The docking of a neurotransmitter to a receptor molecule triggers an electrical excitation that propagates towards the next neuron. Thus, the electrical impulse impinging on the synapse has finally overcome the gap.

Important: The docking of the neurotransmitter causes the receptor molecule to form new. This is because the docked neurotransmitter leads to an enlarged receptor molecule that has a different structure than the original one. This altered structure triggers numerous effects. The most important one is the electrical excitation mentioned above.

We have three problems to solve: Firstly, why the experience of the consciousness network dominates our experience. Secondly, why there is integration and thirdly, why there is differentiation.

We are now looking at a person, let us call him Hugo, who sees a red ball (my favorite example) lying in front of him. Let's simplify the situation and say that Hugo's consciousness network consists of only two modules, the one for "red" and the one for "ball". So he only perceives the red ball, he experiences nothing else.

The modules "red" and "ball" are spatially separated in the brain, but all neurons in both modules fire synchronously, i.e. they always emit their electrical impulses at exactly the same time. Let's say this happens in intervals of 100 milliseconds. As a result, every 100 milliseconds a huge number of receptor molecules is formed simultaneously in their synapses. The modules "red" and "ball" therefore create many simultaneous flashes of consciousness every 100 milliseconds.

As explained, the flashes of consciousness have both a spatial and a non-spatial character. Their spatial character causes them to be located in the brain. Their non-spatial character prevents their separation. Hugo therefore not only experiences "red" and "ball" simultaneously, he also experiences them as a unit, i.e. as a "red ball". This explains the integration, one of the three problems has been solved.

Why does Hugo perceive the red ball continuously? When the concentrated flashes of consciousness occur 100 milliseconds apart. Because he's only perceiving the flashes of consciousness. Gaps between them are inevitably not perceived.

Now Hugo looks at two objects, a red ball and a blue cube. His consciousness network then comprises four modules. As far as the blue cube is concerned, what I just said about the red ball also applies to his integration. But now there is also the aspect of differentiation. Because Hugo experiences the red ball and the blue cube separately. The timed firing of the neurons of the consciousness network is responsible for this. The neurons in the "red" and "ball" modules fire synchronously every 100 milliseconds. The same applies to the "blue" and "cube" modules, which also fire synchronously every 100 milliseconds. But now comes the decisive point: "red and ball" and "blue and cube" fire at different times. Suppose their firing is shifted by 50 milliseconds. If the modules "red" and "ball" fire at exactly noon, then the modules "blue" and "cube" fire 50 milliseconds later, another 50 milliseconds later "red" and "ball" fire again. And so on.

This causes Hugo to experience the red ball and the blue cube at different times, so he perceives them separately. On the other hand, he experiences both continuously, since there is no experience of the gaps. Both aspects together result in the red ball and the blue cube lying next to each other in the experience of Hugo. This solves the problem of differentiation.

The dominance problem remains to be solved. If the consciousness network consisted only of the four modules "red", "ball", "blue" and "cube" firing at intervals of 50 milliseconds, then there would be enough large gaps for the emergence of many other flashes of consciousness caused by formations of molecules outside the consciousness network. In fact, however, a huge number of modules is active in the consciousness network, which causes concentrated flashes of consciousness in extremely short time intervals. Thus all other flashes of consciousness play only a subordinate role.

The consciousness of animals

All forming atoms and molecules have a very short flash of consciousness at the moment of their formation. This of course means that all living beings have consciousness, including bacteria. Because they are full of activity, molecules are formed in huge numbers. But our human consciousness represents a completely different quality of consciousness because it is generated by the temporally coordinated formation of molecules in the brain's consciousness network. So the question arises what the consciousness of the animals that have a brain looks like.

Worms and insects have the most primitive brains. They consist of about one million neurons. Humans have 100 billion, 100,000 times as many. Moreover, it is very questionable whether there is a timing when their neurons fire. Therefore worms and insects have only a very weak consciousness, in no way comparable with our human consciousness.

It becomes more interesting with vertebrates, which include fish, amphibians, reptiles, birds and mammals including humans. Because their brains are all designed similarly. There is the brain stem, which controls all life-supporting functions such as breathing. The cerebellum is responsible for the coordination of the movements. We know that cerebellum and brain stem do not contribute to consciousness. Only the forebrain is important for this because it serves demanding tasks such as planning and decision-making. The forebrain of vertebrates shows the greatest differences. Especially the outer layer of the forebrain, this is the cortex. Only mammals have it and it plays a very important role in the consciousness network. Therefore it can be assumed that all fish, amphibians, reptiles and birds do not have consciousness networks with temporal coordination. Their consciousness is therefore not very pronounced and has no resemblance to our human consciousness.

In humans, the cortex is strongly folded, otherwise, it would not fit into the skull. If it could be smoothed, it would have the area of four A4 sheets. It's four times the size of our closest relative, the chimpanzee. In comparison, the size of the cortex of a rat is that of a stamp. Undoubtedly, chimpanzees also have consciousness networks with timing. But they're a lot smaller. Nevertheless one can say that their consciousness is quite similar to our human consciousness, even if weaker.

Interesting is the question about the consciousness of elephants and whales. Because their brains are considerably larger than the human brain. This is relativized with respect to the cortex: Humans have about 12 billion cortex neurons, elephants and whales (including dolphins) have slightly less, about 11 and 10 billion, respectively. But the difference is small. Then why are we so much smarter? One could first assume that the human cortex neurons are more strongly connected, but this is not the case, the connection density is the same. And yet there is a considerable difference: the speed of the electrical impulses in the connections between the neurons is about three times higher in humans than in elephants and whales.

What does that do? As we have seen, the timing of consciousness networks does not occur immediately, the electrical impulses have to run back and forth several times before the timing has built up. What this means is quite clear: the higher the speed of the electrical impulses, the larger the networks of consciousness can be. And the larger the consciousness networks, the smarter you are. In fact, the intelligence quotient of a human being correlates very strongly with the speed of its electrical impulses. Therefore the consciousness networks of elephants and whales are much smaller than those of humans despite the comparable size of the cortex. Their consciousness is correspondingly weaker. But it is in principle very similar to our consciousness.

Conclusion

Disclaimer: WARNING! If you decide to proceed beyond this point, you must understand you're entering a realm in which few have travelled. If you are familiar with physics, this realm will appear as a Pandora's Box of mystery and intrigue. If you have no formal understandings in physics this section may appear intuitive and almost pragmatically straight forward, but I will introduce a few basic equations.

You do not need to explicitly understand every equation included, but they are provided to support theoretical discussions. At this point, every reader is on the same level. PhD's, researchers, experimentalists and theorists, will all learn this new understanding at the same time.

Do not rely on your previous knowledge to get you through this section, as you are aware, I intend to change everything you think you already know about our universe. For now, refrain from expressing your objections until you reach the bitter end. I promise I will provide you a step-by-step disposition explaining why we should accept these abstract ideas.

Made in the USA
Monee, IL
04 March 2020